Sacred Mountains

Sacred Mountains

A Christian Ethical Approach to Mountaintop Removal

Andrew R. H. Thompson

UNIVERSITY PRESS OF KENTUCKY

Scholarly publisher for the Commonwealth,
serving Bellarmine University, Berea College, Centre College of Kentucky,
Eastern Kentucky University, The Filson Historical Society, Georgetown College,
Kentucky Historical Society, Kentucky State University, Morehead State
University, Murray State University, Northern Kentucky University, Transylvania
University, University of Kentucky, University of Louisville, and Western
Kentucky University.
All rights reserved.

Editorial and Sales Offices: The University Press of Kentucky
663 South Limestone Street, Lexington, Kentucky 40508-4008
www.kentuckypress.com

Library of Congress Cataloging-in-Publication Data

Thompson, Andrew R. H.
  Sacred mountains : a Christian ethical approach to mountaintop removal /
Andrew R.H. Thompson.
      pages cm. -- (Place matters : new directions in Appalachian studies)
  Includes bibliographical references and index.
  ISBN 978-0-8131-6599-8 (hardcover : alk. paper) —
  ISBN 978-0-8131-6601-8 (pdf) — ISBN 978-0-8131-6600-1 (epub)
  1. Ecotheology—Appalachian Region. 2. Mountaintop removal mining--
Appalachian Region. I. Title.
  BT695.5.T474 2015
  261.8'8—dc23   2015035923

This book is printed on acid-free paper meeting
the requirements of the American National Standard
for Permanence in Paper for Printed Library Materials.

Manufactured in the United States of America.

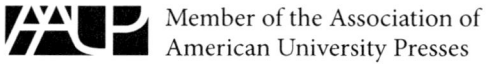

Member of the Association of
American University Presses

They will not hurt or destroy
on all my holy mountain;
for the earth will be full of the knowledge of the Lord
as the waters cover the sea.

—Isaiah 11:9

Contents

Photographs follow page 80

Preface

Ethics in Its Place

To consider place is to consider the particular. To attend to place is to be mindful of the lived, embodied experience of a particular locale. As philosopher Edward Casey argues, knowledge is based on perception, and perception always occurs in a place; in this sense, all knowledge is "local knowledge," in that it arises from embodied experience in a particular place.[1] Thus, to be really useful and accountable to human experiences, our reflections of whatever sort—ethical, philosophical, theological—must literally be grounded; they must attend to places.

This book can be characterized a number of ways. It is a work of ethics. As such, it would typically be described as a work of Christian environmental ethics, although I believe and will argue that its relevance exceeds both these qualifiers. It addresses an issue—mountaintop removal (MTR) coal mining—that is uniquely Appalachian, so it also fits into the genre of Appalachian studies and incorporates many of its disciplines, such as sociology and history. But cutting across these characterizations, and insinuating itself into multiple aspects of my approach, is the issue of place. This is a book about ethics in a particular place.

The notion of place itself receives explicit consideration at multiple points (specifically, chapters 4–6). The idea of a special relationship to place has long been used to characterize the people of Appalachia. As I argue, this association has at times been problematic. Nonetheless, the issue of MTR inevitably and forcefully causes us to pay attention to place, to the particular places that are being radically transformed. It forces us to reflect on how we conceive of places like the Appalachian Mountains—as pristine wilderness? as resources? as home?—and how these conceptions influence us and the mountains themselves.

Thus, an ethical response to MTR simply cannot avoid a consideration of place. Beyond that, the approach I describe, theocentrism, is especially attuned to place because of its emphasis on the concrete and the particular. Throughout this work, I address the details—scientific data, personal stories, popular narratives, local histories—that shape and challenge moral action in this particular place. I articulate an approach that responds to this issue not in the abstract but in the concrete places where it is being practiced and contested. I offer one proposal of how ethics might be done in particular places, even as those places are being transformed in powerfully unprecedented ways.

Introduction

Overturning Mountains

They put their hand to the flinty rock,
and overturn mountains by the roots. . . .
But where is wisdom to be found?
And where is the place of understanding?
—Job 28:9, 12

In April 2008, on a misty morning in the mountains of Kentucky, a group of Christians gathers in the shadow of a surface-mined mountain for a prayer service. The liturgy follows the pattern of the Stations of the Cross; the wooden cross they carry has a large piece of coal secured to it with barbed wire. They pray for the health and renewal of the mountains and mountain communities around them; they mourn the greed and thoughtlessness that are destroying that environment for the sake of cheap energy. In response to a lament of "corporate greed," they intone, "Let us take the symbol of the cross as our sign of generosity." Allen Johnson, the founder of Christians for the Mountains, explains, "The cross shows the victory of God in the face of death and oppression, that the last word is not death, despair, but is hope, life, resurrection."

On another day, on a mountain in southern West Virginia, Andrew Jordon, the owner of Pritchard Mining, convenes a Bible study group in a small cabin overlooking a 1,400-acre surface mining operation (relatively small for this method of mining). The participants discuss what stewardship of creation's resources means in this context. One regular member of the group has been inspired to capture the site in paintings. Jordon is proud of what he has created here. He employs 165 people in an area where work is hard to come by, and he prides himself on training and caring for his young employees. He believes his efforts to reclaim mine sites, some of which were abandoned by previous operators, represent responsible stewardship and real environmentalism.

1

These people and many others like them are drawing on their religious beliefs to respond to the mining practice known as mountaintop removal (MTR). This kind of mining has become a distressingly contentious issue in the Appalachian Mountains, dividing communities and churches and leading to emotional and even violent confrontations. It has been criticized for polluting the land, destroying ecosystems, injuring and sickening families, and even costing jobs, since its reliance on huge machines means it requires fewer and differently trained employees. At the same time, the debate over this practice touches on deeply held beliefs about what it means to be a miner, to love the mountains, and to live in Appalachia, incorporating and manipulating a wide range of assumptions and stereotypes about these ideas. That this issue calls for a response from the church is, as several Christians have put it, "a no-brainer." But contrary to what they might argue, determining an appropriate Christian response to MTR is not so simple.

The purpose of this book is to articulate one such response. I propose an ethical approach that is capable of adequately addressing the culturally freighted beliefs and identities that figure so prominently in this debate. These beliefs and identities must be critiqued and, where necessary, modified, without subtly relying on other stereotypes or assumptions (a weakness that has characterized too many responses to this issue and others like it). Christians and others cannot sufficiently understand this issue, let alone address it responsibly, without carefully examining how it has been constructed on an ideological foundation of discourses and narratives of Appalachian and American identity dating back more than a century.

To articulate such an ethic, I turn to the Christian moral philosophy of H. Richard Niebuhr, who examines historical relativity and the multiplicity of interdependent claims on an agent and interprets them theocentrically, in light of their ultimate source in God. I show that such an ethic is especially well suited to what has been called the "intertextuality" of Appalachia: the intersecting social discourses that shape our understanding of the place and of this issue in particular.[1] My central argument is that a Christian ethical response to MTR requires the church to address the cultural complexity of this debate by reinterpreting its narratives (or, in Niebuhrian terms, its imaginations) with divine purposes, rather than finite human ones, at their center, thereby relativizing all other claims and values. Such a theocentric approach, I argue, can support a radical challenge to the destruction caused by MTR and by the debate surrounding it, but from a theological perspective that is fundamentally different from that of many current responses.

If this turn to a Christian ethical approach seems arbitrary, in a sense, it is. Like Niebuhr, I am convinced there is no neutral perspective, and as theologian Thomas James writes of Niebuhr's position, "one must simply stand where one finds footing and try to make one's position as intelligible as possible."[2] At the same time, the appeal to Niebuhr represents a methodological choice. Willis Jenkins argues for a pragmatic approach to environmental ethics, beginning from concrete problems and the resources communities use to respond to those problems in an effort to clarify and constructively criticize their responses.[3] I draw on Niebuhr's moral anthropology and theology because they provide a useful framework for approaching this task in a fraught, complicated context. The Niebuhrian method I propose here, though described from a Christian perspective, can help clarify and challenge any worldview, even for those who consider the method's substantive theological commitments suspect. While he steadfastly insists that neutrality is an illusion, Niebuhr himself describes his approach using the more universal philosophical terminology of value as well as the particular theological language of revelation. In his ethic of responsibility, he seeks to develop an "instrument of moral analysis which applies to any form of human life including the Christian."[4] Thus, I contend that even for nontheists, the theocentrism I propose offers moral clarity in this complex issue. I develop this claim further in chapter 3.

Before outlining my argument, a brief overview of MTR itself is in order. The focus of my work is not the scientific data but rather the discourses people draw on to frame, and frequently to obscure, those data. Others are better qualified to assess the environmental or economic costs and benefits of the practice (and I cite their work below). My goal, in contrast, is to examine the interpretive frameworks Christians and others use to understand and respond to these data and, where necessary, to offer more appropriate frameworks. Nonetheless, the data matter, especially since (among other reasons) we can adequately understand the debate and discourses surrounding this issue only if we comprehend what is being debated and what is at stake. Therefore, I describe the practice of MTR and note some of its more notorious effects here (these effects are discussed in greater detail in chapter 1). Then I outline the main contours of the debate over MTR, indicating some of the relevant literature as well.

Moving Mountains: Mountaintop Removal Mining

MTR is an extreme form of surface mining practiced in Appalachia.[5] As opposed to underground or deep mining, whereby coal is extracted from beneath the ground with relatively limited disturbance of the surface, surface mining refers to a number of techniques that remove everything covering the coal—rocks, soil, and vegetation, collectively referred to as "overburden"—to access the coal more easily and more completely.[6] Surface mining is more efficient than underground mining, as it involves bigger operations utilizing enormous machines and fewer workers to mine multiple seams at once. It is practiced where underground mining would be unfeasible or uneconomical; in theory, at least, it is used to access seams that are too narrow or too near the surface to be mined by more traditional underground methods. Different forms of surface mining remove overburden in different ways and with varying degrees of surface disturbance; the technique depends on the properties of the coal seams below. MTR, which uses explosives to clear hundreds of feet of overburden from the top of a mountain, is the most radical of these techniques.

MTR begins by removing trees and vegetation from the area to be mined, followed by the removal of topsoil.[7] The trees are frequently logged and sold by timber companies, and the topsoil may be retained for later reclamation. Once part of the land is cleared, a dragline may be brought in and assembled on-site. A dragline is an enormous earthmover, up to twenty stories tall, with a shovel the size of a house that is capable of scooping 100,000 pounds of debris at a time.[8] The dragline is the most infamous symbol of the massive scale of MTR, but its size makes it impractical on many sites; in these situations, more typical equipment (still huge by most standards) is utilized.[9] The scale and mechanization of MTR make it extremely efficient, requiring far fewer workers to mine the same amount of coal by underground or even other surface mining techniques. The exposed overburden that remains is drilled, blasted with explosives, and removed from the area. Some of this overburden is used to regrade previously mined areas. Extra overburden is pushed into adjacent valleys in what are called (often pejoratively) "valley fills."[10] When the coal seam has been exposed, it is broken up by further blasting and hauled away.

After an area has been mined, reclamation begins, frequently while other parts of a site are still being mined. Topsoil and other materials are returned to the area and graded. Since passage of the Surface Mining Control and Reclamation Act (SMCRA) of 1977, sites are required to be returned to their "approximate original contour" (AOC), although this requirement is frequently waived for a variety of postmining uses such as airports, golf courses, and alleged fish and wildlife habitats (some of the more contentious examples of AOC variances are discussed in chapters 4 and 5).[11] Under the SMCRA, mining companies must purchase bonds that are repaid to them after adequate reclamation has been performed; forfeited bonds are used to reclaim the abandoned sites.[12] Far from the haphazard process suggested by many descriptions of MTR, reclamation is often a highly technical endeavor that involves ongoing scientific research and coordination with a number of regulatory agencies.[13] Nonetheless, even some in the industry, such as mine owner Andrew Jordon, acknowledge that much reclamation has been desultory at best.[14]

MTR became widespread in the 1970s. Its use increased through the 1980s, partly, and ironically, because of the SMCRA and its provision for AOC variances, without which the current scale of MTR would be nearly impossible.[15] The practice grew even faster in the 1990s, encouraged, again ironically, by Clean Air Act amendments that increased the demand for the less-polluting low-sulfur coal abundant in the Appalachian Mountains.[16] Although underground mining still accounts for the majority of coal mining in Appalachia, surface mining, including MTR, continues to increase.[17] Reliable details about the extent of MTR are difficult to come by, because of both the contentious nature of the debate and the rate at which new sites are excavated. One survey conducted in 2009 reported that all forms of surface mining had covered nearly 1.2 million acres in Appalachia, representing 10 percent of the central Appalachian region, or 500 mountains affected or destroyed.[18] Based on these numbers, another estimate claimed that by 2012, more than 2,500 square miles would have been mined by MTR and 3,500 miles of streams covered by fills.[19]

Consequences of Mountaintop Removal

This dramatic practice is changing the topography of the oldest mountains in the United States. The Appalachian Mountains are among the most biologically diverse regions in North America and home to several endangered species.[20] A combination of geology (the oldest highlands region in the country), geography (a north-south orientation that supports a high degree of migration), and history (the glaciers of the latest ice age never reached the mountains) has created mixed mesophytic (medium moisture) forests with unparalleled biodiversity.[21] The rivers and streams of the Southern Mountains (another name for the Appalachians) are rich with life. Because the glaciers did not stop the flow of major rivers here, Appalachia's aquatic systems are the most biodiverse and home to the greatest number of endemic species (found only in this region) in the United States.[22] There are more species of fish in Tennessee alone than in all of Europe.[23] In this ecological context, whatever the benefits of MTR, and regardless of the quality of reclamation, the costs—forests cut, streams buried, and mountaintops excavated—are great.

In this process of radical transformation, some of the most widespread and deleterious effects involve the aquatic systems. Even small streams play an important role in nutrient cycling and the production of organic matter, which are crucial to the mountains' food web. Burial of these streams can permanently alter these processes.[24] In areas where more than 5 to 50 percent of the watershed is affected by MTR, water quality and biodiversity suffer, even in streams that are not directly touched by valley fills. Soil compaction and loss of topsoil and vegetation affect runoff in mined areas, leading to flooding, decreased electrical conductivity, and high levels of pollutants, which are then transmitted through the food chain.[25]

Not surprisingly, this kind of ecological impact has consequences for human communities as well. Contact with affected streams, exposure to airborne toxins and dust, and polluted drinking water cause a variety of human health problems.[26] Reduction in the ecological integrity of streams (that is, their ability to support a community of life comparable to that found in similar habitats) in coal mining regions (not just those subjected to MTR) has been linked to increased mortality from a variety of cancers.[27] Similarly, air and water contamination near coal mines and coal processing plants has been associated with an increase in cardiovascular disease and heart attacks.[28] Areas near surface mines also report more hospitalizations for pulmonary disorders, hypertension, lung cancer, and chronic heart, lung, and kidney disease than in other areas.[29] Some of these effects are discussed in greater detail in chapter 1.

Nor are these the most dramatic effects of mining and MTR. Flooding, allegedly increased by the disturbances caused by MTR, has repeatedly devastated coalfield towns. Blasting and floods can damage buildings' foundations and wipe out gardens.[30] When the dam supporting a reservoir of coal waste ruptured in Martin County, Kentucky, in 2000, it spilled 300 million gallons of sludge into nearby streams and communities.[31] A boulder that was dislodged from a surface mine site crashed into one family's trailer home, killing a three-year-old boy.[32] Events like these may be exceptional, and mining is certainly not the only industry to suffer such tragedies. In any case, the environmental, public health, and safety repercussions of MTR are only some of the relevant factors to be considered. Nonetheless, the extent and seriousness of these consequences illustrate that the stakes are very high.

The Debate over Mountaintop Removal

The debate over MTR is wide ranging and highly polarized. Like other similarly contentious issues, even the terminology is contested. Opponents of MTR describe the terms "overburden," "reclamation," and "fill" (rather than "waste") as "Orwellian."[33] Advocates of the practice say the term "valley fills," although first used by the operators themselves, has become misleading.[34] And the industry eschews the term "mountaintop removal," opting for the arguably less descriptive "mountaintop mining."[35]

This contest over terminology is, of course, only one part of a much larger conflict waged in the courts; in local, state, and national government; and sometimes in violent face-to-face confrontations. Opponents of MTR include musicians, artists, and writers; former miners and former mine inspectors; and (most emblematically) local residents who were forced to confront MTR when it directly affected them or their communities.[36] Many are proud of the coal mining heritage in their families or communities but troubled by the effects of MTR. Beyond those effects already described, opponents point to family home places and cemeteries being destroyed and whole communities being forced to leave because of MTR.[37] One West Virginian describes the horror of seeing her grandson standing in a stream full of dead fish, poisoned by mine runoff.[38] The family of a woman from Kentucky battled a mysterious illness for seven years before learning that mining and mine waste had polluted the land and water around their home.[39] Maria Gunnoe, a well-known anti-MTR activist and recipient of the Goldman Environmental Prize for "grassroots environmental heroes,"[40] lists the effects she and others have experienced: "Since mountaintop removal moved into my backyard in 2000, I've lost two access bridges, the use of my water, and five acres of land. I've been flooded seven times. The mountain behind me is crumbling on my home. Everyone downstream from where that mountaintop removal site is gets flooded, and their wells are contaminated. My well is contaminated.

Can't drink my water."[41] Gunnoe's story and others like it motivate people from Appalachia and beyond to oppose MTR.

Meanwhile, miners and others who support the practice have their own motivations. Obviously, these include the fear of losing jobs, especially during a recession in one of the most economically depressed regions of the country. But this fear is about more than economics; for many people in the coalfields, mining is the only thing considered "real work."[42] Yet their motivations go beyond even this. Many families and communities identify closely with the long tradition of the dangerous, difficult, and heroic work of mining. People in this region are proud that they—who have often been dismissed as ignorant hillbillies—are responsible for providing electricity to much of the eastern United States. They see activists and MTR opponents as outsiders (whether geographically or simply in spirit) who are trying to rob them not only of their livelihoods but also of their dignity. Mine manager Randall Maggard chokes up when he describes how his children are taught to be ashamed of his work, arguing that he always strives to do the right thing and is proud of what he does.[43]

These deeply held and highly personal motivations have led to dramatic confrontations between the two sides. When the late Larry Gibson, an anti-MTR activist, held a July Fourth cookout with fellow opponents of MTR, several miners showed up and threatened physical violence.[44] Gibson also reported shots being fired at his home. At a reenactment of the historic Battle of Blair Mountain (which pitted the nascent miners' union against local government and industry thugs in 1921), several unemployed miners, perhaps misunderstanding the significance of the event, harassed and attacked the reenactors, including former US congressman and onetime miners' hero Ken Hechler.[45] At one protest against Massey Energy, speakers, including Episcopal priest Jim Lewis, were shouted down; activist Judy Bonds (another Goldman Prize recipient) was struck by a Massey supporter; and several others, including ninety-four-year-old Hechler and NASA scientist James Hansen, were arrested (as was Bonds's assailant).[46] And at one active MTR site, environmental activists staged a tree-sit, living for several days on platforms in trees slated for removal, enduring threats from mine operators and police, but also reportedly having valuable and meaningful conversations with some of the mine's employees.

Examples like these abound. At the same time, these confrontations do not tell the whole story. The industry has its Don Blankenships (the bellicose former CEO of Massey); it also has its Andrew Jordons (the owner of Pritchard Mining).[47] The opposition has its prophets, voices crying out unequivocally against MTR, in isolation or in concert; it also has its pastors, more moderate figures seeking dialogue. There are moments of tense and even violent confrontation; there are also real moments of mutual understanding. Clearly, emotions run high. There is much at stake. Claims on both sides go beyond concerns about environmental or economic goods: at the heart of this debate are people's lives, safety, history, and identity. How can conflicting claims of such depth be adequately evaluated? Before outlining my own theocentric proposal, I consider other treatments of this and related issues.

Discussions in the Media and Scholarly Works

This debate has received significant attention in the popular media.[48] Most articles focus on policy issues and legal battles.[49] The trope of local grassroots movements confronting the staggering power of mining companies has been a mainstay of these popular works.[50] Some media coverage has focused on the role of religious groups in the debate; many popular books also give attention to the religious aspects of resistance.[51] In general, popular coverage tends to characterize the issue as a central conflict in the debate over energy and environmental policy, pitting environmental and social sustainability against the short-term avarice of an archaic industry. The Appalachian people are important characters in this struggle, filling the various roles of victim, villain, and hero; Appalachia itself is depicted as either pristine wilderness or the home of a backward culture of environmental despoliation.

Academic research on the effects of MTR and other forms of mining emerges from the fields of environmental science, public health, or public policy.[52] Chapter 1 examines much of this research. In general, these studies tend to treat the issues (whether MTR specifically or mining more broadly) in isolation, without addressing the complex cultural background of the Appalachian region. At worst, they trade on some of the same cultural assumptions found in popular accounts.[53] Images of pristine nature degraded by human activity and of impoverished victims oppressed by outside capitalists—indeed, the very notions of culture and nature—have been key symbols in the depiction of Appalachia.[54] This characterization has been created and re-created as a symbolic other, manipulated in the service of the social and political interests of national identity.[55]

Given this context, one in which environmental and political questions are never simply environmental and political, it is crucial to address the assumptions about and images of Appalachia. Recent sociological and anthropological approaches to MTR have sought to interrogate the role of these assumptions in the debate, with varying effectiveness.[56] More generally, scholars in Appalachian studies have worked to examine and change the terms of the debate. Studies have dealt with the images and cultural forces involved in the creation of Appalachia as a region and an identity,[57] as well as with the dynamics and manipulations behind the missionary impulse and development efforts throughout the twentieth century.[58] These concerns have direct implications for understanding the dominance of the coal industry in parts of Appalachia, since it exercises its power in part by controlling the images and symbols that shape consciousness within and beyond the region.[59] Resistance to political and economic domination, particularly but not exclusively that of the coal industry, must therefore address the symbolic and interpretive elements of that domination, as some Appalachian scholars have argued.[60] The standard tropes of grassroots victim-heroes battling corporate greed must be challenged and nuanced. This is necessary not simply to refute them but also to move beyond assumptions and stereotypes and allow the region and its residents to speak for themselves, with all the multivocality this entails.[61]

In this contest of interpretations, religion plays a significant role, as scholars are beginning to realize.[62] To the characterization of mountain religion as primarily otherworldly and socially disengaged (which is not altogether inaccurate), scholars have added reflections on the varieties of religious expression in the Appalachian region, describing religious resources for survival and resistance.[63] Nor are religious responses lacking in the debate over MTR, although research on them remains sparse. Articles in both academic and more popular publications have documented these responses and the motivations behind them from a primarily social-scientific or journalistic (rather than theological) perspective.[64]

In terms of constructive theological work, two pastoral letters written by the Roman Catholic Bishops of Appalachia deal with a broad spectrum of regional social and environmental issues.[65] A subsequent response by the Catholic Committee of Appalachia examines the application of the bishops' arguments.[66] Theologian Michael Iafrate applies a postcolonial theological lens to Appalachia and the problems facing it, including MTR.[67] Finally, sociologist and Appalachian activist Helen Matthews Lewis and theologian Mary Ann Hinsdale combine ethnographic research with contextual theological reflection in their discussion of grassroots organizing efforts in a rural Virginia town.[68] Aside from these texts, little has been done to address environmental and social issues in Appalachia from a theological-ethical perspective. To date, there has been no attempt to bring theological ethics to bear on the specific and urgent moral issue of MTR. With this book, therefore, I seek to fill a regrettable gap in both contextual and environmental theologies (by addressing the Appalachian context specifically) and in Appalachian studies (by considering the religious response from a theological, rather than purely sociological, perspective). My primary goal, however, is to discern and describe an ethical approach that is capable of responding to the complexity of values and identities surrounding MTR with a theology that is both appropriately nuanced and adequately demanding for this urgent task.

A Christian Ethical Response to Mountaintop Removal: A Theocentric Proposal

In this context of conflicting claims and intertwined identities—what I call the intertextuality of Appalachia—the attempt to articulate a consistent Christian response is confronted by what Niebuhr and James M. Gustafson refer to as the multidimensionality of value.[69] With respect to environmental questions (and, in fact, any moral question), values of any sort—streams and communities, pollution and illness—are valued or not valued in the context of their relationships to other beings. The assertion that one particular thing is good (or bad) necessarily raises the question: good (or bad) for whom or what? These are not the abstract relationships of moral reflection but the concrete relationships that frame our interactions with the world. The mountains, for example, are variously valued as homeland, as a place of natural beauty, and as a source of energy. What is more, the same people may value the mountains in *all* these different ways. Therefore, problems that are apparently environmental in nature cannot be adequately considered in isolation from the value-laden, socially constructed discourses that describe them.

For this reason, my approach here turns from the data about MTR and its effects and focuses instead on the images and interpretations that establish and circumscribe those data. I argue that the theological-ethical work of Niebuhr, with its careful attention to the multidimensionality of value and the ways a moral agent responds to multiple claims and identities, provides an ideal Christian ethical approach to this issue. Based on Niebuhr's work and that of other writers influenced by him, I show that an appropriate Christian response to this issue interprets it and the questions surrounding it theocentrically—that is, placing God and God's purposes, rather than finite human interests, at the center of its interpretations.[70] In this approach, the church must assess and reframe its imaginations to reflect these universally inclusive divine purposes. On its surface, this approach may seem quietistic or passive, but I argue that it can establish a strong critique of the worst abuses of MTR.

Throughout the book, I turn to examples that illustrate this approach in communities engaged in responding to this issue. I do this, first, to keep the task of rehabilitating imaginations concrete and accountable to the experiences of those communities. Second, these examples suggest that my approach, though novel, is not a radical departure from the current lived moralities of communities. I am advocating a focus on practices that are capable of fostering new imaginations. Some of these are already part of communities' lives; others may require adapting current practices or inventing new ones.

I show that the complex and value-laden character of the data on MTR necessitates an ethical approach that is capable of interpreting and organizing, rather than reducing, a number of intersecting value claims. Further, I argue that a theocentric approach based on Niebuhr's moral philosophy is more appropriate than other potential environmental ethical perspectives. Accordingly, I begin by addressing, in chapter 1, some of the relevant research on the environmental, economic, and political impacts of MTR in Appalachia. It is undeniable that the practice has significant effects, yet I argue that the data are incomplete and sometimes ambiguous. More important, the way the data are presented and received is invariably shaped by the value systems of those involved. An adequate treatment of this issue therefore requires going beyond the data to the interpretations that circumscribe and define the issue.

This negotiation of values is the concern of ethics. In chapter 2 I turn to environmental ethics and defend the appropriateness of a theocentric approach to this issue, based primarily on Niebuhr's conception of value. I describe several movements within environmental ethics that emphasize the relationships between environmental issues and social dynamics. I then turn to the social construction of value, arguing that none of the other perspectives gives sufficient critical attention to the relational nature of value to adequately address the social complexity of this issue. In contrast, Niebuhr argues that because value is relational, a coherent system of value requires some center around which all other value relationships are organized. For the theist, this center is God. I contend that, although the various environmental ethics all contribute valuable insights to MTR, this attention to the relational nature of value and to how moral agents make sense of conflicting value relationships is what makes a theocentric approach ideally suited to the intertextuality of Appalachia and MTR.

Accordingly, in chapter 3 I turn to a theocentric approach built on Niebuhr's conception of values. The language of value is just one language Niebuhr uses to understand the relationality of moral agents; the language of faith and ethics is, I believe, more representative of his overall approach. Yet, given the multiplicity of disciplines and audiences with which Niebuhr engages, articulating a unified theocentric approach based on his work requires some care. Focusing on his books *Radical Monotheism, The Meaning of Revelation,* and *The Responsible Self,* I outline the fundamentals of a theocentric approach based on the work of Niebuhr and two Niebuhrians, James Gustafson and Emilie Townes. I then describe how this theocentric ethic is applicable to MTR, specifically, from the perspective of the church (which is Niebuhr's perspective as well). My approach involves the reimagination of key issues in the debate over MTR through a process of critically examining and refocusing the church's narratives on the inclusive purposes of God, the universal valuer.

This approach is applied, in chapters 4 and 5, to three specific discursive pairs related to MTR in Appalachia: power and powerlessness, insiders and outsiders, and destruction and reclamation. Narratives about these concepts have long characterized discussions of Appalachia and play an important role in the debate over MTR. When the imaginations of these concepts place finite selves and interests at their center, rather than God and divine purposes, they are, from a theocentric perspective, destructive or evil, leading to alienation rather than responsible moral action. The first step in a theocentric response, therefore, is to examine the role of these narratives in the church's participation in the debate and then to seek out more God-centered images. In chapter 4 I examine each of these pairs in turn, discussing their history in relation to Appalachia and the role they have played in the church's imaginations, with reference to specific statements and practices from a variety of Christians and Christian communities. I argue that although the imaginations of these concepts are not necessarily inaccurate, they are incomplete, neglecting the multivocality and complexity of the realities of Appalachia.

This process of critical examination is revelatory in itself, since it exposes the political nature of these discourses and brings to light the complex realities they often obscure. Nonetheless, the theocentric ethic goes further than this. Imaginations are inadequate not simply because they are incomplete but also, and more importantly, because they are organized around finite human interests rather than divine ones. Accordingly, my approach calls for a reconception of these understandings, with God and divine purposes at their center. How to do this is the focus of chapter 5. Building on the discussion in chapter 4, for each set of concepts I indicate how they have been inadequately imagined or overly limited. I then provide a concise statement of a theocentric understanding of these concepts and explain what that might mean with respect to MTR, citing specific examples of communal practices in Appalachia that incorporate such imaginations.

In the final chapter I place my approach in the context of the common life of the church, at the same time seeking to resolve certain key concerns about this approach. First, I draw together the various strands of my argument to consider this approach to MTR as a whole. Then I place it in the larger context of an overarching theocentric morality. Based primarily on Gustafson's careful explication of Niebuhr's work, I offer a pattern for theocentric moral action: six practices that I believe can integrate the process of critical examination into the church's common life. Here, too, I suggest concrete practices that specify this pattern.

Another relevant aspect of the church's life is its relationship to place. I address this topic by way of a response to a criticism of the theocentric approach. This ethic, and its application to this issue, may be seen as too passive or conciliatory, with its emphases on discernment and humility rather than specific norms or clear ethical boundaries, and on divine action rather than human action. Some might argue that this kind of approach limits the possibilities for outspoken criticism of unjust practices. Certainly in light of MTR's cost in human and nonhuman lives and communities, such a charge needs to be taken very seriously. I respond to this concern in three ways, arguing, (1) that this approach actually strengthens, rather than weakens, the possibilities for prophetic critique; (2) that attention to the experience of a place, characterized by the notion of "loving the mountains," can be incorporated into this approach; and (3) that in light of the notion of an Anthropocene epoch, theocentrism has good reason to denounce MTR not as wrong in itself but as symptomatic of a broader pathology of human-centered imagination. With these elements, my theocentric approach can indeed support a strong criticism of unjust practices, although it does so on a radically different theological basis from other approaches to this issue. I conclude by describing some of the general moral guidelines about MTR that I believe might be part of this criticism. I place these guidelines at the end of the study because I believe they are the outcome of, not the foundation for, careful discernment.

This critical perspective notwithstanding, my approach is perhaps more moderate than many in this highly charged debate. I argue for a critical examination of the narratives that guide the debate, more open and inclusive dialogue, and greater attention to the details of MTR. To the extent possible, I strive to give equal consideration to the arguments of both sides. Most of the critical scrutiny, however, is directed toward the opposition to MTR. This is not because my sympathies lie with those who support the practice or because I believe they are not guilty of similar distortions. On the contrary, it is precisely because of the well-documented and troubling consequences of MTR, as well as the divisiveness of the debate, that a discerning and critical Christian response is so urgent. Because the most vocal and developed Christian responses to date have been in opposition to MTR, these receive the bulk of attention, negative and positive, in this book, with the hope of not simply criticizing them but of clarifying and strengthening them.

I have great respect for many of the Christians on both sides of this debate whose faith has compelled them to respond to this issue. Many of them have devoted significant thought and prayer to the challenges posed by MTR. It is out of this respect that I offer my proposal. My goal is to advance this process of discernment of a Christian response that takes seriously the rich cultural and historical context in which the debate over MTR is situated. The stakes are high. Lives and livelihoods, homes and identities are threatened. Communities are being divided and century-old loyalties dissolved. Mountains are overturned, yet understanding still eludes us. An appropriate Christian response, one that understands these issues in light of God's ongoing purposes in creation, is urgently needed.

1

Downstream Impacts

Environmental, Economic, and Social Effects of Mountaintop Removal

The most logical place to begin a discussion of MTR and the debate surrounding it would presumably be the environmental sciences and what they reveal about the practice's effects on ecological systems. It is not a foregone conclusion, however, that ethical deliberation would necessarily proceed from an examination of the realities of a particular moral problem. As Willis Jenkins points out, the choice of such an approach is a methodological one, representing a pragmatic strategy as opposed to a "cosmological" one, which would begin by clarifying a fundamental worldview and its attendant moral commitments before applying the latter to particular situations.[1] In later chapters (especially chapters 3 and 6) I adopt such a pragmatic method, arguing that the particular details of MTR are an important part of my theocentric approach. At this point, however, my decision to begin with these details has more to do with the overall goal of this work. One of the claims I defend throughout the book is that too much discussion of Appalachia and its challenges is based on assumptions and stereotypes, and that the main problem with these assumptions is not their falsity but their incompleteness. To offer a more complete representation of Appalachia, I must begin with a clear and comprehensive description of the issue at hand.[2]

As discussed in the introduction, the impact of MTR on the surrounding ecosystems and communities is dramatic and devastating. Despite the seemingly self-evident nature of such destruction, determining the type and extent of MTR's effects on the environment of Appalachia is complicated and involves a number of interacting systems and scientific disciplines. In this chapter, I present some of the available research on MTR's influence on biological systems, including human communities. To provide as complete a picture as possible, this inquiry must also consider the economic and political contexts in which MTR operates. In addition to the catastrophic effects on the environment, MTR has (by at least one reliable estimation) a negative net economic impact on the states where it operates, and it has been the subject of numerous contentious court battles and policy reversals.

The apparent certainty of these conclusions is misleading, however. The data on MTR are incomplete, and more research is needed. More important, the research that is available is often contested or denied. As I argue in this and later chapters, the presentation and interpretation of the data are shaped by value and how the interested parties value the relevant factors. Thus, these deliberations need to be guided by an ethical approach capable of negotiating the value relationships involved in this issue. First, however, I turn to the data.

Aquatic Systems

Much of the research on MTR has focused on how valley fills affect aquatic systems, for two main reasons. First, the practice of dumping overburden in valleys and thereby burying streams falls under the jurisdiction of the Clean Water Act, and many have seen this legislation as a promising means of combating the practice. Second, the effect of valley fills on streams can be quite dramatic and can extend far beyond the immediate vicinity of the mine because small headwater streams flow into larger bodies of water. In some cases, the effects of MTR have reached the wells of nearby communities, polluting families' tap water.

DIRECT LOSS

The most obvious impact of valley fills on aquatic systems is, of course, the burial of hundreds of miles of streams by overburden. According to an Environmental Protection Agency (EPA) study that covered approximately 12 million acres in southern West Virginia, eastern Kentucky, southwestern Virginia, and a few counties in northern Tennessee, 724 miles of streams were buried by valley fills between 1985 and 2001. A later study that included other mining activities, such as blasting and road construction, showed that more than 1,200 miles of stream were buried by mining between 1992 and 2002. More recent statistics are unavailable, but the EPA estimated that stream losses would double between 2002 and 2012.[3]

The small headwater streams lost to valley fills house biological communities that are different from those of larger downstream waters. When these streams are buried, the rich biodiversity of the Appalachian region is compromised. Many of the species that inhabit headwaters are listed as endangered, threatened, or of special concern.[4] When valley fills kill these animals and destroy or fragment their habitats, they also harm the surviving members of the species by making the population as a whole less resilient and more susceptible to environmental changes.

These small headwater streams also perform unique ecological functions that have a significant impact on downstream ecosystems. These streams remove or transform contaminants, such as nitrogen; hold sediment and woody debris, keeping it from traveling downstream; and process large volumes of leaf litter and organic matter, providing energy for the stream's food web.[5] When streams are buried, these functions are effectively lost.

Defenders of MTR have argued that many of the affected streams are ephemeral or intermittent, running only in certain seasons or only after rainfall (as opposed to perennial streams, which run nearly continuously), and therefore of little importance. On the contrary, the EPA argues that even after long drought periods, these intermittent streams can support diverse, balanced, and unique biological communities.[6]

INDIRECT IMPACTS

In addition to the loss of the biological communities and ecological functions of headwaters, several other factors contribute to the downstream impact of MTR: the clear-cutting of trees and other vegetation on the mine site; the collection of surface and groundwater within the valley fill itself; and the compaction of the surface by heavy equipment, which increases run-off.[7] These effects can alter the flow of a stream. For example, by absorbing water, the valley fill moderates stream flow, keeping a stream flowing during droughts and reducing flow during low-intensity rainstorms. After intense storms, however, streams below valley fills exhibit higher flood levels than other streams; in some cases, the streams had flood levels that would be expected to occur naturally only every 50 to 100 years, compared with 10 to 25 years in unmined areas.[8] The higher proportion of sediment in streams below valley fills, partly due to the loss of the headwaters' sediment-collecting function, exacerbates this flooding. Obviously, this is a serious problem for nearby communities. Regulations have been implemented to reduce these increases in flooding, and research suggests that alternative ways of placing valley fills may mitigate this problem.

The chemical composition of streams below valley fills is also altered. The most troubling difference is an increase in selenium,[9] which has been associated with reproductive failure, gross deformities, and death in fish, birds, and other animals in stream ecosystems. These streams also have, on average, ten times the concentration of chemical ions of streams in unmined areas, as indicated by water conductivity.[10] These chemical changes lead to an appreciable decline in biodiversity among invertebrates, fish, and salamanders. In structure, function, and biodiversity, streams below valley fill sites are significantly degraded.[11]

RECLAMATION

Mitigation of the effects of valley fills typically involves creating channels to restore at least some of the functions of the buried stream. However, because of the significant structural differences between a channel and a stream, these channels are generally unable to reproduce the water quality and biological diversity of headwaters.[12] Although these channels perform some of the same functions as unaffected streams, such as the breakdown of leaf litter, their structure and biology are sufficiently different for EPA researchers to argue that they "[do] not adequately replace natural ephemeral [streams]" and "should not be considered as onsite mitigation for the natural channels buried under [valley fills]."[13] More recent reclamation efforts have sought to construct channels that better mimic the characteristics of natural streams and better replicate their function and biology; so far, however, it is not clear how a constructed channel might be able to change the drastically altered water chemistry caused by valley fills, including elevated conductivity and selenium levels. Without improvements in water quality, it is unlikely that biological communities can be restored.[14] Similarly, restoration of the riparian forest below a valley fill can help prevent erosion and provide inputs of leaf litter and wood, but these efforts are unlikely to improve the water chemistry. The only effective way to minimize the chemical changes resulting from MTR is through dilution from nonaffected streams.[15]

Terrestrial Impacts

Although much of the research and litigation surrounding MTR has focused on its impact on stream ecosystems, the practice has significant consequences for terrestrial ecosystems as well.

LOSS OF MOUNTAINTOPS AND FORESTS

Most obviously, MTR removes mountaintops, altering the topography of the region. According to the US Geological Survey, mountaintops were lowered an average of 34 meters, and valleys were raised by 53 meters.[16] This topographic transformation affects other aspects of the mountain ecosystem, including vegetation, soil, biodiversity, and even climate. The surface temperature at mine sites is higher on average and shows greater variability than in nonmined areas, at least partly because of the lower elevation. Moreover, air cools as it rises up a mountain, and the moisture in it begins to condense, leading to what is called orographic precipitation. Lower mountaintops may result in less orographic precipitation.[17]

Even before the mountaintops are lowered, MTR affects mountain ecosystems by removing forests from mine sites. The mixed mesophytic (moderate moisture) forests that dominate this region are unique, combining both northern and southern species.[18] They have been described as "the most biologically diverse ecosystem in the southeastern United States."[19] Thousands of acres of these mixed mesophytic forests, along with the oak and pine forests that also grow in Appalachia, are cut down by MTR operations, and the trees are usually just dumped into a valley fill. The destruction of these forests devastates valuable habitat for Appalachian species, removes a primary source of energy production (that is, energy production by plants), and eliminates an important form of protection against erosion.[20]

The destruction of thousands of acres of forest also transforms the remaining forest ecosystems through fragmentation. As more and more sections of forest are destroyed, the proportion of the remaining forest that is considered edge forest increases, and the proportion considered interior forest decreases. A 2007 study showed that the amount of interior forest lost due to MTR was actually 1.5 to 5 times the amount of forest directly destroyed by MTR.[21] This is important because of the significant differences between the two types of forest. Edge forests are warmer and drier than interior forests, their trees and animals are more vulnerable, they have more nonforest and invasive species, and they have higher concentrations of pollutants. Many species, particularly songbirds like the cerulean warbler, require interior forest habitats.[22]

Finally, the removal of mountaintops and forests leads to the loss of a great deal of topsoil, which has significant repercussions for reclamation efforts. Although regulations require that topsoil be salvaged and redistributed for postmining land use, in practice, exemptions are frequently granted because the topsoil is presumed to be too poor or too thin to be worth saving. Indeed, these claims may be true: soil that is well suited to hardwood forests may not be appropriate for the grasslands that are typically the focus of reclamation efforts.[23] Thus, rather than being saved for reclamation, topsoil is often discarded with the rest of the overburden.

LOSS OF BIODIVERSITY

In the mixed mesophytic forests of Appalachia, the number of endemic species (those occurring only in this region) is high. Of the 351 vertebrate species in Appalachia, 14 are found nowhere else in the world; 9 of these species are salamanders.[24] The salamander fauna of the Appalachian Mountains is one of the richest in the world; according to one conservative estimate, there are about 10,000 salamanders per hectare of forest.[25] Because of Appalachia's distinctive geographic location, the region is home to mammals suited to northern, temperate, and tropical climates. The avian population of the mixed mesophytic forests is one of the richest in the eastern United States.[26]

The biodiversity of the region also includes the forests themselves. In the West Virginia portion of the previously mentioned EPA study area, accounting for about 2.8 million acres, there are ten different types of forest.[27] MTR disproportionately affects higher-elevation forests, such as the northern hardwoods that grow on the upper slopes of mountains; valley fills are likely to have disproportionate effects on cove forests, which grow in headwater settings.[28]

By altering the distribution of forests, MTR shifts the balance of wildlife communities. What were once mature interior forests, such as the high-elevation hardwood forests and headwater cove forests, become edge forests or grasslands. Accordingly, prevalent animal species shift from interior forest species to grassland and edge forest species. Bird species that require large tracts of mature forest are reduced in mined areas. Among amphibians and reptiles, there is a shift from salamanders to snakes and other reptiles. This is obviously a serious concern. At the same time, defenders of MTR point out that the new habitats created by the practice, including ponds as well as edge forest and grassland, have had a positive impact on several bird species, including the wild turkey, bobwhite quail, and ruffed grouse. Small mammals actually become more abundant in reclaimed grassland areas. The effects of MTR on large mammals have not been studied, but there is anecdotal evidence that it has a positive impact on white-tailed deer.[29] Breeding populations of some species previously considered rare or unknown in this region have been documented on reclaimed mine sites.[30] It is worth noting, however, that the previous rarity of these species was most likely due to the rarity of their preferred habitats, not necessarily because these species were endangered or threatened; many are common in other parts of the country.

FOREST LOSS AND CLIMATE CHANGE

One of the most prominent environmental consequences of MTR is its relationship to climate change, which is multifaceted. Most straightforwardly, MTR replaces a carbon sink with a carbon source. That is, it destroys a mature forest capable of sequestering carbon dioxide from the atmosphere in order to access stored carbon in the form of coal, which will eventually be released as carbon dioxide into the atmosphere. In addition, the mining activities themselves—land clearing, excavation, transportation—burn fossil fuels and release carbon dioxide. For these reasons, many people concerned about climate change see MTR as one of the most egregious offenders. Studies suggest that even after forests have been reestablished, reclaimed sites sequester carbon at a lower rate than undisturbed forests.[31]

Climate change may also interact with MTR to intensify its impact on animal species.[32] As average temperatures increase, many animals move either toward the poles or to higher altitudes. Obviously, a mountaintop represents the upper limit of this latter migration; by lowering the mountaintops, MTR reduces these species' ability to adapt to changes in temperature. Some salamander species are projected to lose significant amounts of temperature-appropriate habitat as early as 2020.[33]

RECLAMATION

As in aquatic ecosystems, reclamation can minimize the negative effects of MTR on forests. So far, however, reclamation has generally been unsuccessful at reproducing a premining forest community.[34] Ironically, the Surface Mining Control and Reclamation Act is partly responsible for this poor record. Prior to its passage in 1977, most surface-mined land was reclaimed with trees, and reforestation created commercially valuable forests. In an effort to reduce erosion and improve the stability of reclaimed sites, however, the SMCRA prescribed heavily compacted, intensely graded sites that were unable to support reforestation and were therefore converted to grassland.

More recently, researchers have described the "forestry reclamation approach," a method that creates loosely compacted sites that are capable of supporting productive native forests, offering habitat for forest species, and providing valuable crop trees for landowners and communities.[35] Moreover, this approach actually costs mine operators less than traditional reclamation practices, due to less intensive grading and less expensive seed mixtures. Regulations in Virginia, Kentucky, and West Virginia support this practice, and mine operators are increasingly opting for forestry as a postmining land use. Nonetheless, there is still a large amount of land that was reclaimed using traditional methods or not reclaimed at all.

Community Health

Some of the most controversial and troubling aspects of MTR are its consequences for human communities. However, precise data on the health effects of MTR are limited. Thus far, much of the available information is anecdotal (like the stories cited in the introduction); too few investigators are focusing on these questions to provide sufficient empirical research. Moreover, many of the studies have addressed the health effects of mining in general, without distinguishing among underground mining, surface mining in general, and MTR in particular. MTR has been linked to a number of negative health outcomes, but causality is difficult to prove. Nonetheless, studies have shown correlations between living in MTR areas and higher mortality and morbidity rates, even after adjusting for other variables.[36]

Humans are affected by MTR's impact on water and air. In the rural coalfields, many people get their household water from wells. Wells near mine sites have higher levels of mine-related chemicals than wells in unmined areas.[37] Families who live near slurry impoundments—huge storage ponds for the liquid waste from coal preparation—have reported diarrhea, rashes, changes to teeth, and kidney stones.[38] Homes in coal-producing counties also have high concentrations of hydrogen sulfide, a gas produced when bacteria come into contact with sulfates, which can be found in mine runoff. Regular inhalation of hydrogen sulfide can lead to headache, irritability, and poor memory. Coal dust is another potential hazard, especially near coal processing facilities. Although no specific tests have been conducted, researchers believe that exposure to coal dust may lead to cardiovascular and lung disease and possibly cancer.[39]

Environmental scientist Nathaniel Hitt and Michael Hendryx, a professor at West Virginia University and a leading researcher into the health effects of MTR, have shown a correlation between low stream integrity in mining areas, based on the invertebrate species living in the stream, and increased rates of digestive, respiratory, breast, and urinary cancers, after controlling for other variables.[40] In another study, Hendryx and colleague Laura Esch found higher mortality rates from cardiovascular disease in mining areas compared with nonmining areas, and higher rates in MTR areas compared with underground mining areas.[41] Another study by Melissa Ahern found that the incidence of birth defects in MTR mining areas is 26 percent higher than in nonmining areas (and significantly higher than in underground mining areas). Moreover, the overall prevalence of birth defects has increased as the practice of MTR has spread (although the prevalence of some particular defects has decreased).[42]

Perhaps most useful is a study by Hendryx and Ahern that examined mortality rates in the coalfields, comparing high-mining areas, low-mining areas, areas of Appalachia without mining, and the rest of the nation.[43] The authors found higher mortality rates in mining areas than in other parts of Appalachia and in the nation as a whole, and these disparities increased over time. The authors then estimated the cost of this higher mortality based on the value of statistical life (VSL) measurement used by US government agencies. They found that the cost of higher mortality in the coal mining regions of Appalachia exceeded the economic benefit of coal as estimated by the coal industry. This was the case after controlling for other variables and using two different VSL estimates. Again, this study does not prove causality. Nonetheless, the overwhelming implication is that "coal generates inexpensive electricity, but not as inexpensive as the price signals indicate because those prices do not include the costs to human health and productivity, and the costs of natural resource destruction."[44]

The Economics of Mountaintop Removal

Almost without exception, defenders of MTR cite its economic benefits as sufficient justification for the practice, regardless of the social and environmental costs. This claim is frequently dismissed as industry propaganda. Opponents point out that even as coal production has increased, mining jobs overall have declined significantly, from around 55,000 in 1975 to 20,000 today, due to increased scale and mechanization.[45] MTR is seen as the biggest reason for this decline, since an MTR mine employing 50 workers can produce as much coal as an underground mine employing 150. In addition, mining companies are often accused of bringing in workers from other parts of the country. Given all these factors, opponents argue that industry claims that "coal means jobs" ring false. But the economic realities of mining are more complicated than this.

There is great demand for Appalachian bituminous coal, both domestically and internationally. This coal is high in thermal efficiency and low in impurities such as sulfur. Since the Clean Air Act Amendments of 1977 and 1990, it has been sought after for electricity production because it produces less pollution than other types of coal.[46] Concerns about energy independence have also increased the demand for domestic coal as an energy source. At the same time, some of the coal in Appalachia is used in metallurgical processes, which gives it a higher value in global markets.[47] Meanwhile, when natural gas and oil become more expensive, coal becomes a more attractive alternative. For all these reasons, the price of coal has risen significantly, doubling in a relatively short period.[48] Thus, it has become profitable to mine seams of coal that previously would have been left unmined.

Increased demand notwithstanding, Appalachian coal must compete with western and international coal. Western coal is generally comparable to Appalachian coal in terms of impurities; it is also closer to the surface and therefore easier and cheaper to mine. This competition, combined with the growing demand for cheap energy, means that Appalachian mines must take advantage of economies of scale to be profitable, mining bigger and bigger sites with increasingly larger and more numerous machines to keep costs as low as possible. In this context, MTR may be the only economically feasible way to mine some seams of coal.

Supporters argue that the coal industry benefits the states of Appalachia through the collection of taxes from those directly and indirectly employed by the industry. According to a study conducted by the West Virginia Center on Budget and Policy and the environmental consulting firm Downstream Strategies, direct employment from mining accounted for 3 percent of employment in West Virginia in 2009, while indirect employment accounted for 9 percent. In coal-producing counties, the average for direct employment was 5 percent, although in some counties it was as high as 47 percent.[49] The average annual wage for coal industry employees was $74,110; the average for miners was lower, but it was still higher than the state average for all wage earners. Based on these data, the authors concluded that direct coal employment accounted for $125.5 million in tax revenue in 2009, and indirect employment accounted for $167.9 million. However, the state also spent money on these employees, in the form of public services and infrastructure. When these expenditures were factored into the calculation, the net economic impact of direct coal employment was about zero, while indirect employment resulted in a net loss to the state of approximately $116.9 million. In other words, based on tax revenue, coal cost West Virginia more than it contributed.[50]

Coal also contributes directly to the state budget through taxes and fees, but as with employment, there are expenses associated with the industry, in the form of administration and road maintenance. In 2009 the coal industry contributed approximately $307.3 million in tax revenue, accounting for about 8 percent of the General Revenue Fund in West Virginia and less than 1 percent of the State Road Fund. The state's costs, incurred to maintain coal roads and run agencies such as the Departments of Commerce and Environmental Protection, amounted to $113.7 million. Thus, coal's net direct contribution to the state budget in 2009 was $193.6 million. Forgone revenue in the form of tax credits and exemptions, however, amounted to $173.8 million. When this was factored in, the net gain to the state was $19.7 million. Combined with the net loss of $116.9 million from direct and indirect employment, the net impact of coal on West Virginia's state budget was a loss of $97.5 million.[51]

Like Hendryx's estimate of the cost of increased mortality due to coal mining, this study is a useful benchmark. However, it does not resolve the question of whether mining is good or bad for Appalachia or West Virginia, even on purely financial terms. To address that question, one would have to acknowledge, for example, that the revenue lost due to tax exemptions for the industry would also be lost if the industry no longer existed; thus, considering this an expenditure can be misleading. Likewise, and more significantly, the costs of providing public services and maintaining infrastructure would remain the same if the coal industry shut down, unless former employees moved away, while the state would lose the revenue it receives as a result of their employment. Put simply, although it may be true that coal costs West Virginia approximately $97.5 million a year, getting rid of coal might not save the state $97.5 million.[52]

Assessing the veracity of the "coal means jobs" argument in all its forms also requires the examination of a simpler but no less significant reality. Whatever the net impact of mining in terms of state revenue and expenditures or in terms of jobs lost or gained by a particular form of mining, on a county or community level, when a particular site is no longer profitable or feasible, for whatever reason, jobs are lost, and that matters tremendously to the people who find themselves unemployed. To be sure, because coal is a nonrenewable resource, those jobs are destined to be lost eventually; but whether they are lost now because the environmental, social, and economic costs are deemed to be too great, or at some distant time when accessing the coal is no longer feasible, makes a difference. Moreover, the communities affected are in some of the poorest counties in the poorest states in the country. The question of whether coal is responsible for creating or alleviating that poverty (likely both), while important, is not the most urgent one for those who depend on the industry. Unlike defenders of MTR, I do not consider this the only, or even the most significant, argument. I do not assume that Appalachia has no economic alternatives to coal; indeed, as I articulate later, finding viable alternatives is an important part of the approach I propose. Nonetheless, I note the local importance of these jobs for two reasons: first, because it helps account for the tenacity and forcefulness of the "coal means jobs" mantra, and second, because it reminds us that the reality represented by economic data and employment figures is not the only relevant reality.

Mountaintop Removal and Policy

One study of MTR from a policy perspective notes, with no apparent irony, that decisions about MTR take place in a "well-established, and continuously evolving" regulatory framework.[53] This description captures both the unwieldiness and the fluidity of mining and MTR policy, which continues to shift in response to environmental litigation and political demands.

LEGISLATION

The key federal laws governing MTR are the Surface Mining Control and Reclamation Act of 1977 and the Clean Water Act (CWA), particularly sections 402 and 404. The SMCRA establishes the criteria for permitting, regulating, and reclaiming surface mines, under the Office of Surface Mining Reclamation and Enforcement (OSM). The CWA "regulates the discharge of pollutants into the waters of the United States."[54] Section 402 of the CWA specifically regulates pollutants or waste and allows the EPA to issue permits. Section 404 regulates fill material in waterways. This provision of the law is administered by the Army Corps of Engineers, which can issue either individual permits for discharges with "more than-minimal impacts," requiring a significant amount of documentation and review, or general permits for discharges with "minimal impacts."[55] Thus, which regulations apply to

MTR is determined by whether overburden is considered waste or fill and, in the latter case, whether it is deemed to have more than a minimal impact. In practice, most MTR sites have been issued general permits according to CWA section 404.[56]

THE CHANGING REGULATORY LANDSCAPE

In 1998 the West Virginia Highlands Conservancy brought a lawsuit against the West Virginia Department of Environmental Protection and the Army Corps of Engineers, alleging that valley fills violated the CWA. In *Bragg v. Robertson,* the plaintiffs argued that overburden was clearly waste, rather than fill, which the CWA defines as material used for "the primary purpose of replacing an aquatic area with dry land or changing the bottom elevation of a water body."[57] The case also alleged that valley fills violated the SMCRA's prohibition against disturbing a 100-foot "buffer zone" around streams.[58] Federal district court judge Charles Haden II agreed with the plaintiffs. Supporters of the industry argued that coal mining could not continue if the verdict stood. Ultimately, an appeal supported by the Clinton administration led to the verdict being overturned by the Fourth Circuit Court of Appeals.

Bragg v. Robertson had two direct outcomes. First, the Army Corps of Engineers and the EPA proposed a rule to clarify the meaning of fill to specifically include valley fills, ensuring that all MTR permit decisions would be made by the Corps of Engineers under CWA section 404. The rule was ultimately adopted by the Bush administration in 2002.[59] Second, the final settlement required the preparation of a programmatic environmental impact statement for MTR; although this statement details the environmental impacts of valley fills, it abstains from any discussion of whether they should be limited.[60]

Repeated legislative and judicial efforts to change the definition of fill material to exclude valley fills have been unsuccessful.[61] In 2003 Judge Haden ruled again that MTR violated the CWA, but his verdict was again overturned by the Fourth Circuit Court of Appeals. In another lawsuit, the Ohio Valley Environmental Coalition argued that the use of a general permit for MTR is illegitimate, since such a permit assumes that environmental impacts are minimal. The plaintiffs alleged that MTR causes significant impacts and should be subject to the more stringent individual permitting process. The court agreed with the plaintiffs and ordered the Corps of Engineers to revoke eleven previously approved permits. This ruling was also overturned by the Fourth Circuit Court of Appeals. Another decision by the district court would have required the Corps to prepare individual environmental impact statements for every new MTR permit, but once again, the Fourth Circuit overturned the lower court.

Other changes have similarly eased the regulatory restrictions on MTR. In 2004 the OSM set out to clarify the stream buffer rule, acknowledging that strict enforcement would effectively eliminate MTR.[62] The new regulation was approved by the Bush administration in 2008. The sanctions for violating the CWA and SMCRA were also reduced. Many of these new rules have been put forward ostensibly to reduce redundancy among the OSM, EPA, Corps of Engineers, and US Fish and Wildlife Service.

In general, the policy context surrounding MTR has been controversial and complicated. According to John Craynon of the Appalachian Research Initiative for Environmental Science, MTR has been characterized by "a rapid movement towards internal regulation and responsible decision-making."[63] Craynon and his coauthors argue that the current status quo is inadequate because government and industry decision-making bodies operate independently of each other and, more problematically, without broad public participation. At stake in environmental issues are "discrepancies between various types of value placed on natural systems."[64] The complex regulatory framework of MTR and the highly contested background that has shaped it, they argue, require an approach capable of negotiating these conflicting values. As a solution, they propose public ecology, which they define as "the nexus of science, engineering, public policy and interest, citizen views and values, market forces, and environmental protection statutes and regulations, which, through an open and participatory discourse, is intended to ensure that the ecological systems continue to function as societies operate within and derive benefits from them."[65] This approach, the authors suggest, may be capable of overcoming the various challenges confronting any discussion of MTR, including its history, scale, polarizing legal framework, and barriers to public involvement, and it may enable real and productive deliberation about data, policies, and values.[66]

Data and Value

From one perspective, the shifting policy framework of MTR, with multiple unsuccessful attempts at regulation and a gradual move toward self-regulation, seems to illustrate only hypocrisy and doublespeak.[67] Less cynically, and more constructively, it can be seen as an illustration of how values can shape and organize the data that define an issue like MTR. As Craynon and his coauthors point out, "science and policy have always been tournaments of value."[68] When the economic context is included with environmental and policy research, this competition of values becomes even more obvious. Data do not speak for themselves. The assumption, voiced by advocates on either side of this issue, that more complete data or more adequate communication of those data will lead to real understanding—that agreement is possible if those on the other side could only see beyond ideologies to the actual facts of MTR—proves false as factual claims are questioned, contextualized, or outright denied.[69] For example, biodiversity may be seen as a relative good. Whether it is acceptable to create habitat for one species at the cost of another species' habitat is an open question, however. That MTR represents a net loss for West Virginia is a matter for empirical investigation (although, as we have seen, even such an investigation admits of multiple interpretations). Yet that may be of little importance to a community whose employment depends on a nearby mine. At the base of this failure of trust and communication, then, are the unarticulated systems of value that shape how data are presented and interpreted. In the absence of an approach capable of articulating, organizing, and negotiating these competing values, distrust and frustration will continue to prevail in the debate over MTR, as in any other complex and freighted debate.

The approach that Craynon and colleagues propose, public ecology, is a promising one. They suggest that public ecology emerges precisely in response to three intersecting needs: the need for communities and local knowledge to address policy concerns, the need for increased dialogue across cultural and disciplinary boundaries, and the need for a larger vision of healthy relationships capable of guiding deliberation.[70] Public ecology calls attention to the ways knowledge is produced across disciplinary boundaries and to the implicit normativity of all scientific inquiry.[71] As a procedural framework for bringing together the various interested parties and seeking to negotiate the values at work, such an approach seems well suited to the fraught context of MTR.

As a procedural framework, however, public ecology is unable to provide guidelines for organizing or prioritizing the values that shape an issue. This approach offers content in the form of scientific or "biocultural" knowledge: a clearer awareness of the precise ways social and environmental forces interact.[72] It does not attend to values themselves, beyond arguing that they need to be negotiated in open, public deliberation. To complement public ecology's salutary insistence on such openness and accessibility, we need an approach capable of guiding this negotiation by naming and assessing the different values involved. With such an approach—with a more thorough understanding of how values shape and interpret an issue, and with a coherent and more or less organized system of values—we can return to the presentation and deliberation of scientific data and political dynamics with more clarity.

Environmental ethics can be considered a collection of attempts to do just that: describe and guide the negotiation of values with respect to environmental questions. In the next chapter, I discuss a number of different environmental ethical perspectives that focus on what the proponents of public ecology call bioculturalism, the interaction between human social dynamics and nonhuman environmental dynamics. Like public ecology, each of these perspectives makes important contributions to the negotiation of values, and to this negotiation in the specific context of MTR. Yet each approach leaves important questions unanswered, particularly about its own foundational assumptions. It is in this respect that the theocentric approach described by H. Richard Niebuhr is most helpful. By organizing all values around a divine center, theocentrism challenges all finite values without resorting to complete relativism. This turn toward value is not a turn away from data; indeed, the theocentric approach insists that scientific research is one way of understanding God's action in the world. Thus, attention to these data is one of the main implications of this approach (see chapter 6). Yet before the data can be considered properly, the value systems that circumscribe them must be named and negotiated. It is to these value systems, and some of the environmental ethical perspectives that seek to negotiate them, that we now turn.

2

Environmental Ethics and the Construction of Values

Scientific data do not speak for themselves; data are inseparable from the value relationships that shape how they are presented and interpreted. Responding morally to an issue like MTR requires critical attention to people's value systems, and these systems are shaped by narratives and discourses about the identity and history of the region and its people. One way of understanding the task of environmental ethics is this: ethical theories are mobilized to name and clarify the values that are or should be applied in interpreting and responding to relevant environmental data. The challenge for a Christian ethical response to MTR, therefore, is to identify an approach that gives sufficient attention to the constitutive role played by these intersecting narratives and images in constructing and defining the issue itself. Failing to do so risks perpetuating the very discourses that have distorted and manipulated perceptions of life in Appalachia, thereby undermining the agency and self-definition of its inhabitants.

In this chapter I argue that the theocentrism of H. Richard Niebuhr, especially his understanding of value, is uniquely helpful in addressing this problem of value construction and negotiation. First, I assay several promising alternative ethical perspectives as they apply to MTR and show that, although they offer important insights, none is sufficiently attuned to its own methodological presuppositions to allow Appalachian intertextuality to "speak for itself." The perspectives I consider—ecofeminism, liberation theology, environmental justice, environmental pragmatism, and political ecology—are promising, in that they all seek, on some level, to understand and address the power of the social discourses that define an ostensibly environmental issue like MTR.[1] Obviously, the treatment of each is brief, and this selection of approaches is neither exhaustive nor even representative of contemporary environmental ethics. In fact, I chose these particular perspectives because they share a significant insight with the approach I adopt—specifically, their implicit or explicit attention to the social construction and relational nature of values. Each therefore offers important contributions to a response to MTR, and two in particular—pragmatism and political ecology—inform the approach I describe. Nonetheless, all are inadequate because they begin from certain foundational assumptions that mitigate their ability to analyze critically the social construction of the issue at hand. Second, I consider each perspective's strengths and weaknesses in relation to MTR and Appalachia in particular. I then turn more directly to the question of values to examine this inadequacy more clearly. I argue that Niebuhr's relational theory of value provides the most accurate and helpful foundation for an ethical approach to MTR, one that is capable of relativizing fundamental assumptions and thereby permitting a thorough critique of the discourses that surround the issue.

Environmental Ethics: Five Perspectives

The most promising environmental perspectives for analyzing how values are constructed around an issue like MTR are those that recognize the inexorable social and cultural forces that shape human valuing. Such perspectives arise in response—and, in a sense, in opposition—to a key tension in the history of environmental ethics, that between anthropocentric and nonanthropocentric foundations for valuing nonhuman nature.[2] Anthropocentric perspectives value the nonhuman world instrumentally, based on its usefulness to human life, while nonanthropocentric ones seek to establish some basis for its intrinsic value. Variations on and hybrids of these two poles abound, yet this question of anthropocentrism has remained a central concern for environmental ethics.

Several more recent environmental ethical perspectives shift the focus away from this question by showing that our valuation of both humans and nonhumans is inseparable and equally distorted—that social injustice and environmental degradation are symptoms of society's moral pathology with regard to what we ought to value and how. In one way or another, each of the five perspectives I discuss in this chapter articulates some version of this basic intuition, and each contributes a particular insight to an understanding of MTR in its multivalent social context. At the same time, each perspective manifests certain limitations that an adequate ethical response to the issue must overcome.

ECOFEMINISM

Ecofeminism begins with a concern for women's place in society and explores how women are uniquely harmed by environmental problems.[3] Although explanations of this harm vary, a central claim of ecofeminist authors is that both social injustices and global environmental crises are rooted in the "mutually reinforcing oppressions of humans and the natural world."[4] At the heart of these oppressions is a dualistic view of nature and culture. Ecofeminist authors challenge this view, which they attribute variously to patriarchal religion, the scientific mind-set, psychology, and Western philosophy. The appropriate response to this oppressive dualism is integrative, approaching environmental and social injustices together and explaining them with reference to the web of interacting ideologies—racism, sexism, classism, speciesism—that shape them.[5] In this effort, some ecofeminists apply the insights of ecosystem ecology, an antifoundationalist approach to ecology that emphasizes holism and complementarity in an ecological network.[6]

In the quest for integrative alternatives to a dualistic worldview, feminist theology has made significant contributions. In particular, theologian Rosemary Radford Ruether has been influential from the earliest days of ecofeminist thought.[7] In her seminal work *New Woman, New Earth*, Ruether articulates a clear indictment of the religious (and particularly Christian) roots of the twin pathologies of sexism and environmental exploitation. More recently, in *Gaia and God*, she examines the cultural beliefs and value systems that have been used to justify such a destructive worldview.[8] She explores various cultural myths and narratives, both religious and scientific, to show how they contribute to ideologies of domination and to discern the resources they have to offer a more holistic culture. She seeks healing both for and from these mythological and narrative resources. Ideologies of domination, she claims, are rooted principally in dualisms that separate humans from nature, spirit from body, men from women, and one ethnic or cultural group from another to justify the subjugation of, exploitation of, and violence toward one by the other.[9] This domination has been paradigmatically acted out in an adversarial view of the female body. Dualistic creation stories, theories of evil, and eschatologies have engendered social structures built on the domination of women and others by male elites. These social systems, today projected on a global scale, underlie the interrelated crises of overpopulation, hunger, depletion of fossil fuels, pollution, extinction, and war, with poverty as a key theme in all these issues.[10] "The same systems of power," she states, "that allow a small percentage of the world's population to monopolize most of its resources also throw the growing masses into conditions of misery and into an environmentally destructive relation to their habitat."[11]

Based on her investigation of the cosmological and ideological roots of contemporary destructive worldviews, Ruether, like many other holistic ecologists, offers as a corrective the ecological concept of Gaia—a vision of the world as one organism, with integrating networks of interdependency and cooperation.[12] She maintains that ethics should appropriate and refine this vision into a holism capable of healing the imagined dualisms of fact and value, private and social, theory and practice. Science can be normatively incorporated into this perspective, describing the complex interrelation and kinship that bind humans inextricably to the rest of nature. Ethically, this worldview calls for conversion to more life-sustaining practices, recognition of sin as primarily a distortion of relationships among humans and with the rest of creation, and a move from the dynamics of domination and subjugation to "biophilic mutuality."[13]

The appreciable influence of ecofeminist ideas on religious reflection about social and environmental issues in Appalachia is discussed below. Ecofeminism's fundamental insight—that ecological crises (including the destructiveness of MTR) and the oppression of women and other groups arise from the same pathological dualism—calls attention to the social dynamics and exercises of power that determine how issues are framed. Moreover, the search for integrative cosmologies and images (like the Gaia image) capable of supporting healed relationships has much in common with the holism of Niebuhr's theocentric approach. At the same time, as I argue below, ecofeminism has certain limitations that a theocentric approach may help resolve.

LIBERATION THEOLOGY

The work of some liberation theologians in relation to environmental crises has much in common with ecofeminism. Liberation theologians in general argue that the poor are uniquely privileged, both as the object of divine concern and as a locus for divine revelation (and, consequently, as an epistemic starting point for ethics). Whereas ecofeminists approach environmental problems from the vantage point of their effects on women in society, liberation theologians begin from the perspective of the poor and oppressed of the world, viewing nature itself as a victim of oppression.[14] In particular, they challenge views that would assign both the bulk of the blame for environmental degradation and the bulk of the responsibility for addressing it to the global poor. From this perspective, liberation theologians and ecofeminists reach similar conclusions, attributing the oppression of both the poor and the earth to a dominant ideology that is divisive rather than integrative, and seeking the solution in new, holistic worldviews.

Leonardo Boff approaches his work with the conviction that "the logic that exploits classes and subjects people to the interests of a few rich and powerful countries is the same as the logic that devastates the Earth and plunders its wealth, showing no solidarity with the rest of humankind and future generations."[15] He attributes the predominant paradigm of domination and power to a loss of connectedness that is fundamentally rooted in anthropocentrism (most often, androcentrism).[16] The hubris that places the entire universe at the service of man[17] severs relationships with others and with nature and drives a never-ending quest for power for its own sake. The earth and its people are reduced to capital in the service of a logic of accumulation and exploitation.[18] According to Boff, this dysfunction and its myth of growth underlie capitalism itself. In contrast to this system of domination, Boff, like Ruether, points to an emerging paradigm symbolized by Gaia.[19] This vision gives rise to an ethic of connectedness and solidarity with the vast community of beings of which humans are a part, rooted in spirituality and a sense of the sacredness of all life, including human life.[20]

For Boff and others, the starting point for this ethic of connectedness is liberation. Given the systemic nature of the paradigm of domination, real solutions must begin from a radically new perspective. They must begin with a social ecology that hears the cry of the poor and the cry of the earth as the same cry. These theologians maintain a preferential place for the poor as the foundation for understanding the interconnected crises facing society. At the same time, the meaning of liberation must be broadened to recognize that all human beings, including the powerful, are oppressed by the paradigm of accumulation and domination and that the earth itself is first among the oppressed.[21]

In this perspective, the exploitative culture of division and domination is understood theologically as sin, humanity's self-assertion that rebels against its own finitude. The remedy is a reconnectedness (*re-ligação,* a play on the word "religion") founded on a covenant with God through grace.[22] The covenant with Noah symbolizes this communion with all creation, and St. Francis of Assisi is presented as an example of the spirituality that supports it. For Francis, communion with creation from the vantage point of poverty was a way of life.[23] According to Boff, such an "inner ecology" of spiritual connectedness, beginning in community with the poor, is necessary to challenge the violence that threatens the outer ecology.[24]

Thus, like ecofeminism, liberation theology insists that the divisive ideologies of anthropo-, andro-, and ethnocentrism, which are responsible for the consumption of both people and the earth, can be countered only by an integrative and holistic mind-set. For liberationists, this mind-set must begin "from below," from the convicting cry of the poor and of the earth. Rather than being in opposition, this perspective argues, the social and environmental consequences of a practice like MTR are mutually reinforcing; they are symptoms of the same pathology. As I discuss below, this understanding, like that of ecofeminism, has already had a significant influence on many Christians' view of MTR and Appalachia, and it provides valuable insight into a Christian response, even as it reveals certain limitations to be overcome.

ENVIRONMENTAL JUSTICE

Environmental justice shares Boff's and others' conviction that environmental and social oppression are two faces of a pervasive system of domination. But in contrast to the more philosophically or theologically based perspectives of liberation theology and ecofeminism, environmental justice begins with the lived, bodily experience of that oppression. Beginning in the 1980s, grassroots groups—led primarily by women of color in urban environments—began to see the real and pernicious relationship between their communities' social and economic marginality and their subjection to a number of environmental harms. Environmental burdens such as pollution, noxious development, and resource depletion were—and are—placed disproportionately on minority communities, which often lack the political power to challenge this maldistribution and, because of scarce economic opportunities, may be more willing to sacrifice environmental and public health for jobs.[25] This inequality is referred to as environmental racism. In response, these groups mobilized in a variety of ways, calling for a more just distribution of environmental costs and benefits.

Churches and religious groups have been integral to this activism, and religious ethicist Willis Jenkins calls environmental justice "perhaps the most significant Christian contribution to public environmental deliberation in the United States."[26] When the United Church of Christ published a report detailing environmental racism in 1987, it signaled an important change in conceptual approaches to social and environmental issues.[27] With the support of church groups, the efforts of grassroots community organizations to combat the symbiosis of environmental harms and social and political marginality have now become a broader political environmental strategy and a challenge to churches, policy makers, and mainstream environmentalists. As a political strategy, environmental justice advocates participatory democracy and decentralized political power as the means to understand and address the complex ways that race, gender, and class affect the distribution of environmental harms.

In keeping with this history, environmental justice is a practical strategy first; only secondarily is it a theoretical framework. Environmental justice begins with strategic action to oppose environmentally and socially discriminatory policies, and that action leads to new ways of understanding injustice. It differs in this respect from ecofeminism and liberation theology (at least in Boff's work), which focus more on worldviews and the conceptual connections between environmental and social injustices. This distinction should not be drawn too clearly, though, because the strategic action of environmental justice certainly draws on its own cosmological and conceptual foundations, and authors like Boff and Ruether look to concrete examples of activism to illustrate and inform their conceptual frameworks. Furthermore, scientist and philosopher Kristin Shrader-Frechette has extensively documented the theoretical principles of the environmental justice movement.[28] Nevertheless, more than the other approaches already discussed, environmental justice is primarily a "problem-focused coping strategy" with roots in social activism and the civil rights movement.[29]

The environmental justice movement has been central to religious and secular opposition to MTR. Individuals' and communities' struggles with mining companies and political bodies have received significant attention.[30] These grassroots struggles have been explored through a variety of theoretical and ethnographic lenses.[31] Additionally, from a religious perspective, environmental justice, like ecofeminism and liberation theology, has had an influence through Roman Catholic pastorals (discussed in more detail below).

ENVIRONMENTAL PRAGMATISM

Like environmental justice advocates, environmental pragmatists begin with a concern for the lived experience of environmental problems and policies. They proceed, however, from this concrete experience to a direct challenge of the theoretical bases of mainstream environmental ethics.[32] They apply the insights of philosophical pragmatism as explicated by Charles Sanders Peirce, William James, and John Dewey to environmental ethical issues. According to environmental pragmatists, ethical approaches that attempt to establish metaphysical foundations for environmental ethics and to answer questions about intrinsic and instrumental value ignore the social processes by which environmental problems are constructed as such, and they misunderstand the way human beings actually value their surroundings. The typical search for a fundamental intrinsic value (whether human or not) on which to base an assessment of conflicting values is, they say, question-begging: it assumes the existence of such a value. Indeed, environmental pragmatists question whether such a foundational value would ultimately matter for moral and political decisions.[33] Instead, pragmatists advocate an understanding of value that is pluralist, antifoundationalist, and relational—a web rather than a chain.[34] Environmental ethics then becomes a matter of describing specific values as we experience them—*this* experience of wilderness, for example, rather than some abstract wilderness *as such*—and assessing and comparing them.[35]

Like the original pragmatists, environmental pragmatists' main concern is for a practical, viable framework for understanding and strengthening concrete participatory political action. A major weakness of the "typical" approach, with its debate over abstract concepts such as intrinsic and instrumental value, anthropocentrism and biocentrism, and monism and pluralism, is its inability to offer concrete solutions to "real world environmental dilemmas."[36] Accordingly, these writers, like their turn-of-the-century philosophical predecessors, appeal to empirical study, experience, experimentation, and deliberative discourse rather than the prescription of one "correct" decision based on a single set of principles.[37]

Applying these insights to religious ethical approaches, ethicist Anna L. Peterson argues that if such approaches are to effectively address environmental problems, they ought to start with ethnographic research. Ethicists should examine how people develop values and choose among them, and they should contribute to the creation of rituals and practices that can help transform these values.[38] Similarly, Jenkins argues that religious ethicists have good reason to take pragmatists' concerns seriously. Their criticisms, he says, show that "the religious ethicist cannot merely supply religious resources to a shared project and then convince religious constituencies to adopt its results. . . . The moral for Christian ethics is to focus less on the ecological quality of worldviews and more on the possibilities within Christian experience for participatory adaptations to contextual problems."[39] We might do this, he suggests, by generating theological responses based on the efforts already being made by faith communities to address environmental problems.

In spite of this endorsement of an ethical theory that begins and ends with practical engagement, Jenkins points out some of pragmatism's limitations. Chief among them is that, by attaching environmental reflection to concrete issues in specific contexts—a move that is necessary to focus pragmatism's otherwise unwieldy pluralism—this approach forestalls debate about how these issues are framed and especially about the power dynamics involved in that framing.[40] What constitutes an environmental issue, what makes it problematic, and what goals are an appropriate response? Who makes these determinations? Thus, while the pragmatists' concern with empirical observation and their pluralistic, relational view of value are useful for reflecting on MTR and for the theocentric approach, their analysis of broader dynamics is limited. Consequently, I turn to a final perspective—or, more accurately, a group of perspectives—that explicitly attends to these broader dynamics in the social construction of environmental issues.

POLITICAL ECOLOGY

What the four previously described approaches to environmental issues have in common is a turn from questions about anthropocentric versus non-anthropocentric sources of value toward a critique of society's inability to properly value both fellow human beings and nonhuman nature. In different ways, each of these approaches challenges the *anthropos* at the center of this standard debate: the debate assumes a monolithic concept of "the human" that is either an appropriate center of intrinsic value or not. The reality is that this *anthropos* obscures myriad relations of power and value, including but not limited to humanity's relation to nonhuman nature. Moreover, these relations are constantly being constructed and transformed by political and social dynamics and structures. The very terms of debate—concepts such as environment, nature, and justice; problems such as destruction, degradation, and pollution; and goals such as reclamation, conservation, management, and wholeness—are already part of power-laden social relations in which some people benefit at the expense of others. Therefore, much is at stake in the framing of issues. These relationships and the construction of notions of environmental value are the subject of political ecology.

Political ecology comprises a wide variety of perspectives, methods, and styles. Indeed, some question whether political ecology has enough theoretical coherence to be described as a single field. Some argue that it is best characterized as a set of tools for examining and challenging the forces at work in the political negotiation of environmental issues and for documenting alternative strategies for that negotiation.[41] Accordingly, works in political ecology are primarily case studies of the social and political discourses surrounding particular environmental issues, such as soil erosion or water pollution, in specific regional contexts (typically outside the United States and western Europe).[42] In spite of this diversity, Paul Robbins (who, by his own admission, is *not* a political ecologist but is a helpful expositor nonetheless) argues that, from a functionalist perspective, political ecology does indeed have coherence as "something people do," and he and other authors have attempted to define it accordingly.[43]

A central insight of political ecology is that environmental change is always a product of political processes, since the effects of change are distributed in ways that are shaped by, and in turn shape, unequal power relationships.[44] In this, political ecology can be contrasted with "apolitical ecologies"—ecological approaches that ignore the political operations behind distribution and development in global society. But by ignoring these political dynamics, "apolitical" approaches are, in fact, political, since they are mobilized within unequal dynamics of power and influence the distribution of environmental goods. Their supposed political neutrality only attempts to obscure their politicized nature. The goal of political ecology, therefore, is to make explicit the political processes inherent in all aspects of human interaction with the environment—the ways "political power flows through ecological systems."[45]

What distinguishes political ecology's approach from, for example, environmental justice is its concern for the construction and application of ecological ideas and narratives informed by post-structuralist discourse theory.[46] Authors challenge the use of ideas such as ecological health and integrity, within a discourse of development, to support a narrative that assumes environmental degradation as a given and attributes it to the destructive behaviors of the inhabitants of the global South. Political ecologists examine the social processes that construct such narratives and look to local narratives and practices for alternative strategies.

The value of political ecology with respect to MTR is clear. An ethical response to MTR must attend to precisely these sorts of politically constructed images and narratives. Like the other approaches discussed, political ecology affirms that environmental problems are bound up with social dynamics; but more than any of the others, it carefully scrutinizes the discursive processes and power-laden relationships through which those problems are framed as such. For this reason, the work of political ecologists is especially useful for both examining religious ethical responses to MTR and articulating my own approach, which emphasizes the theocentric reimagination of these discourses and relationships.

Environmental Ethics and Mountaintop Removal

Since my goal is to articulate a Christian ethical approach that adequately attends to the complex cultural discourses surrounding MTR, it is worth considering how some of the above-mentioned perspectives have been or might be mobilized to respond to this issue and the social and environmental questions related to it. This serves several purposes: it makes the previous discussion of the different environmental ethical perspectives more concrete, it focuses on the issue at hand (MTR), it allows us to explore the strengths and weaknesses of the various approaches in relation to MTR, and it highlights some specific concerns that any ethical response must consider. Although each of these environmental ethical perspectives provides valuable insights into the appropriate response to MTR, a consideration of their shortcomings is equally helpful, since it suggests the need for an approach that is more attentive to the intertwined images and narratives—the intertextuality—of this issue, which, as I argue, can be met by turning to the work of H. Richard Niebuhr.

ECOFEMINISM AND LIBERATION THEOLOGY: ROMAN CATHOLIC PASTORALS

Ecofeminism and liberation theology figure prominently in church documents pertaining to mining and MTR. Two pastoral letters from the Catholic Bishops of Appalachia, for example, clearly reflect prominent aspects of these two perspectives (although the bishops would probably not describe their position as ecofeminist). These pastorals, which are framed as a response to "the cry of Appalachia's poor," address economic and societal ills in general, but the focus inevitably turns to mining, including surface mining and MTR.

The influence of liberation theology is unmistakable in the first letter, *This Land Is Home to Me*, written in 1975. The suffering of the Appalachian people, which includes poverty, unemployment, unsafe working conditions, and lack of education and effective support programs, is attributed to "corporate giants," to a social structure that makes the poor dependent and turns them against one another, and to a cultural ethos of consumption and maximization of profit that sets itself as an idol.[47] Although structural sin is not referred to explicitly, the economic system is seen as a force "accountable to no one," a system in which "it is hard for good people to do good," yet one that is ultimately dependent on individual actions: "Economics is made by people . . . those who claim that they are prisoners of the laws of economics only testify that they are prisoners of the idol."[48] This characterization echoes liberation theologians' emphasis on the structural or systemic nature of sin. In the pastorals, this system of sin is contrasted with the life of the poor, the "children of the mountains," whose suffering is a symbol of suffering around the world and whose struggle is to realize the dream of a "life free and simple, with time for one another, and for people's needs, based on the dignity of the human person, at one with nature's beauty, crowned by poetry."[49] Finally, a faithful response to the injustice and oppression of Appalachia requires mutual action to empower the poor to liberate themselves. A new social order is being born, and the church must join in working toward a just society.

In spite of these similarities with liberation theology, *This Land Is Home to Me* does little to trace the mutually reinforcing cultural dynamics that support social and environmental injustice the way Boff and the ecofeminists do. The second pastoral, *At Home in the Web of Life*, written in 1995, shifts the emphasis from social injustice and oppression to the violation of the unity of creation, giving it more in common with these other perspectives. Like its predecessor, this letter sees the suffering of Appalachia as a symbol of the suffering of the wider society, and it attributes this suffering to industry and postindustrial consumerism; behind these societal ills, according to the bishops, lies a "culture of death." Here, however, and in common with the analyses of Boff and Ruether, the culture of death is what violates and divides the ecological whole of creation, turning humans against nature and against one another and reducing them to mere resources. The solution is a culture of life "in which people and land are woven together as part of Earth's vibrant creativity," evoking the ecofeminists' and liberationists' central image of Gaia.[50] Women play a special (if not necessarily prominent) role in this pastoral, as both uniquely oppressed and central to the creation of the new culture.[51] And like both Ruether and Boff, the bishops call for sustainability rather than development, mutuality and community rather than domination, and the nurturing of a spirituality of connectedness and loving relationships.

These letters articulate a compelling call for justice for the people and land of Appalachia. Even though they are hortatory works meant for a wide audience and are therefore more interested in persuasiveness than in analysis and nuance, they are characterized by the careful social and historical examination and theological reflection typical of the Catholic social thought tradition. They offer a moving, poetic appeal for a new culture of life, and they populate that culture with rich theological and ecological imagery. These pastorals have been influential in the social ministries of the church,[52] and I return to them elsewhere in this book. However, their relatively uncritical reliance on stereotypical images of the mountains and of mountain people as simple, slow, compassionate, and connected to nature—the exact opposite of modern consumer culture—is problematic.[53] Nor are the pastorals unique in this respect. Other liberation- and feminist-influenced approaches to Appalachia leave many of these romanticized visions of the mountains and their inhabitants more or less unscrutinized.[54] These approaches presuppose a narrative framework of oppression and liberation, and their proponents find—perhaps more in the myths and images of Appalachia than in its realities—this narrative replicated in struggles such as the one involving MTR. This view fails to recognize the great variety and internal complexity of social and environmental conflict. Regardless of the similarities that undoubtedly exist between contexts, the too-easy application of a narrative of liberation obscures important—and morally relevant—differences.

As scholars like Allen Batteau, Henry Shapiro, and Stephen Hanna argue, even such romantic, apparently positive images of Appalachia manipulate notions of the region to fit preconceived assumptions and political interests. The main problem with such images is not their accuracy or inaccuracy; the issue is that these images are presented independent of the social and political contexts from which they arise.[55] For instance, if the goal is to articulate an ethical approach to MTR that is capable of addressing the stereotypes that surround it, one should not begin with "the dream of the mountains' struggle," however compelling that image might be.[56] These perspectives have significant strategic value in their ability to link environmental issues, including MTR, to struggles for social justice and wholeness, and some elements of their narratives and images may be self-consciously mobilized to great effect. Even so, without careful attention to the political construction of these images, such efforts remain, in Batteau's words, "suspended in webs of significance they have not spun."[57] They are unable to provide the critical tools for articulating, in a richly complex cultural context, precisely what notions of wholeness and justice might mean. More-over, as I argue later, from a theocentric perspective, these perspectives risk "absolutizing the relative," placing finite human values, rather than inclu-sive divine ones, at the center of moral reasoning. If a theological-ethical approach to MTR is to take this complexity seriously—as it must, to avoid further stereotyping and manipulating Appalachia and its people—it needs to incorporate the urgency and holistic thinking of ecofeminism and lib-eration theology without appropriating their uncritical application of the narrative of liberation.

ENVIRONMENTAL JUSTICE: GRASSROOTS MOVEMENTS

In addition to ecofeminism and liberation theology, the pastorals from the Catholic bishops express clear affinities with environmental justice. This is particularly true of the solutions proposed by *This Land Is Home to Me*. These center on processes of "listening to our people, especially the poor," with the goal of citizen and community control over economic and political struc-tures.[58] The bishops also note, without further discussion, the importance of "community organizing and citizen control," as well as "public voice in local, state, and national politics."[59] This call for more participatory demo-cratic structures in response to systemic social and environmental injustice is characteristic of the environmental justice perspective.

Well beyond the pastorals, however, environmental justice has argu-ably influenced the MTR debate more than any other perspective, not only those considered in this chapter. Stories of grassroots community organi-zations have dominated the portrayal of this issue in the popular concep-tion. Anthropologist Bryan McNeil, for example, recounts the experiences of two of these groups: Coal River Mountain Watch (CRMW) and Friends of the Mountains (FOM). He argues that in their opposition to MTR, these groups reimagine notions of community, economy, environment, and moral-ity in ways that run directly counter to the neoliberal ideologies expressed by MTR.[60] According to McNeil, activists in CRMW and FOM center these alternative conceptions around the idea of the commons, rooted in their own interpretation of Appalachian history and culture, to create a new kind of environmental activism that addresses issues more comprehensively than the dominant "myths" of either corporate environmentalism or bourgeois environmentalism, both of which exclude human activities and interests from their concern for the environment.[61] He believes that the environmen-tal justice activism of these groups reveals that healthy economic regimes require a "workable concept of justice" and social structures (like CRMW and FOM) capable of organizing communities' interests "in opposition to powerful industrial and political interests."[62] McNeil's analysis illuminates the ideological resources being mobilized (primarily unconsciously, one assumes) by grassroots groups operating within an environmental justice framework.

Nonetheless, environmental justice perspectives on MTR, as described by McNeil and others, face limitations similar to those of ecofeminism and liberation theology as expressed in the pastorals. Like the narrative of oppression and liberation used by those approaches, categories of justice and injustice provide a ready framework for action, but one that may obscure the complexities of MTR and Appalachia. Nor do McNeil's more nuanced categories of a historically rooted commons environmentalism versus the dominant narratives of neoliberalism escape this risk. While he portrays the grassroots groups' ideology as one that rejects the dualisms of corporate or bourgeois environmentalism, he establishes his own dualism wherein the groups' sense of justice and their appreciation of the commons are opposed to the myths that separate humans from the environment.[63] The activists' reimagination of categories of community, economy, and environment based on the idea of the commons may be seen as a strategic and selective mobilization for their own political purposes, but McNeil does nothing to interrogate the assumptions or problems inherent in this choice. Meanwhile, his use of the idea of myth with reference to the narratives of industry and bourgeois environmentalists, but not to the narratives of the grassroots groups, reinforces the sense that the latter's narratives more accurately represent the demands of justice.

Other examinations of environmental justice perspectives on MTR suffer from similar shortcomings.[64] Stories of groups like CRMW and FOM illustrate the significant strategic value of environmental justice as a framework for understanding and, more importantly, for acting when environmental harm is unequally distributed and communities' lives and dignity are at stake; this includes MTR. I have no desire to diminish that strategic importance. Yet when environmental justice, like ecofeminism and liberation theology, fits MTR into preestablished categories, it neglects the social complexity and cultural multivocality of the region and its people. If, as I have argued, an appropriate ethical response must attend to this intertextuality, the myths and imaginations employed by environmental justice groups require more serious critical scrutiny than McNeil and others offer.

ENVIRONMENTAL PRAGMATISM: ETHNOGRAPHY AND EMPOWERMENT

A pragmatic approach to MTR faces problems similar to those of ecofeminism, liberation theology, and environmental justice. As Jenkins notes, when the narrow scope of pragmatist inquiry is itself interrogated as to its premises, there is no way to respond without recourse to the very meta-ethical debates pragmatism seeks to avoid.[65] One example of a pragmatist approach to an issue similar to MTR was a project to gather ethnographic data about Appalachians' relationship to their land.[66] These data were used to inform the state government of the potential impacts of a proposed power line on residents.[67] The main author of the report admitted the ambiguity of this criterion: although the research was indeed empowering, in that it made the residents feel important, their actual impact on decisions about the proposed power line was unclear. Moreover, the project failed to examine the main criterion of citizen empowerment and prescinded from the debate over the relationship between empowerment and other potential values. Nor did it address the multifarious power dynamics involved in the process of conducting research into subjects' emotions and expressions and presenting those findings to a state body for ruling. My point here is not that the project was flawed but that it was limited, which, as Jenkins points out, pragmatist projects must be.[68] Although it scrupulously avoids problematic abstract values such as justice or liberation by appealing instead to empirical research, environmental pragmatism must mobilize that research in the service of other unquestioned values, be they empowerment, management, or sustainability.

When pressed further, some pragmatists turn to a kind of intuitionism, invoking the sense of value in nature that is present, even if minimally, in everyone's immediate experience.[69] Common ground based on this experienced value can then be the foundation for practical negotiation between different sets of values. Even if the pragmatists are correct to begin ethical deliberation from persons' irreducibly concrete experiences of value (and I believe they are, as the discussion of the experience of place in chapter 6 makes clear), the ability to reach an acceptable compromise in any but the most superficial cases is far from certain. Moreover, the appeal to "immediate" experience assumes that one's sense of the value of nature—a particular landscape, creature, activity—is unaffected by the social and political influences that shape every part of one's perception. As political ecologists would aver, every experience of nature (and even the concept of nature itself) is already politically fraught. To disregard the political dynamics that influence and shape our experiences of value is to limit the scope of our moral reasoning. While pragmatism's suspicion of the search for fundamental unifying values and its attendant concern for irreducible experiences of value are salutary in articulating a response to MTR, its inability to challenge its own premises takes too much for granted.[70]

POLITICAL ECOLOGY: NARRATIVES OF THE MOUNTAINS

What, then, can political ecology, with its critical attention to precisely these concerns about the construction of environmental issues, offer? Several authors have taken this approach to MTR and other Appalachian issues, analyzing the political and social forces at work in the creation and interaction of our experiences of value in nature. One study looks at the different conceptions of the commons by two groups—ginseng harvesters and MTR mine operators—with respect to the same mountain spaces.[71] The author argues that the community's narratives of ginseng, the most economically valuable renewable resource in Appalachia, constitute an alternative conception of the commons that is not dictated by narratives of progress and development, and that these ought to have a central place in the imagination of a postcoal economy for the region.[72] Another study examines the coal industry's ideological construction of a coal-centered image of Appalachia, through such strategies as the creation of a faux grassroots group called the Friends of Coal.[73] These articles follow the long tradition of Appalachian studies scholarship that has scrutinized the construction and manipulation of Appalachian culture and identity since the nineteenth century. Studies of this sort are essential for dissecting the intersecting influences operative in the social construction of environmental issues. What political ecology is unable to do, however, is assess normatively the values invoked in these issues. As a set of tools for examining the construction of issues and documenting alternative strategies, political ecology is more descriptive than prescriptive; whatever prescriptive solutions it offers center on more inclusive democratic processes rather than substantive norms.[74] The ginseng study can offer an alternative construction of the commons; it cannot, with consistency, normatively recommend that alternative. The study of ideology construction can point to inconsistencies in the coal industry's narrative of Appalachian identity, but it is ill equipped to tell us why those inconsistencies matter.

　　Other political ecologists deconstruct pragmatist-endorsed values such as conservation, management, sustainability, and development as tools of normalization and control.[75] Yet to move beyond this critical deconstruction and offer conclusions, these authors, much like the pragmatists, appeal either explicitly or implicitly to notions of emancipation, resistance, or local knowledge.[76] Alternatively, they use the study to offer a more complete theoretical understanding of ideological construction, as the Friends of Coal study does, without any normative assessment. Whereas the other approaches are guilty of fitting MTR (and other issues) into predefined normative frameworks of justice or liberation, political ecology begins—and therefore ends—with little if any normative framework at all. This is to be expected of a post-structural critical approach; it is, however, inadequate for purposes of this study. The approach I describe and apply incorporates political ecology's valuable scrutiny of politically constructed discourses within a normatively theocentric worldview.

Environmental Ethics and the Construction of Values

The environmental ethical approaches discussed thus far share a sense of the relatedness of ostensibly environmental issues and social ones or, indeed, the permeability of the distinction between the two. It is this awareness that makes these approaches useful in articulating a religious ethical response to MTR. This may seem like no more than a superficial similarity; after all, the environmental justice movement's action-oriented political consciousness seems quite different from the political ecologists' deconstruction of narratives of management or sustainability. Yet this connection between environmental and social issues rests, in all these cases, on a recognition that the way we define and delimit environmental issues, what we include and what we exclude, is of great significance. Further, these perspectives remind us that this process of definition is propelled and shaped by power-laden social and political dynamics that must be examined and, at times, challenged. Both in this shared attention to the social construction of environmental issues and in their shortcomings and lacunae, these approaches indicate the importance of understanding values and how they are derived and interrelated.

VALUES IN THE FIVE PERSPECTIVES

Each of the five perspectives points to specific ways that human interaction with the environment is defined and shaped by social and political dynamics. Liberationists and ecofeminists argue that dualisms—man and woman, culture and nature, and mind and body, among many others—and the consumption-driven culture of domination founded on such dualisms blind society to the radical integrity of life on earth, such that even efforts at sustainable development fail to address the real social pathology of loss of connection. Environmental justice challenges the dominant narratives of a democracy where fairness—that is, distributive and procedural justice—is promised but profit and power dominate. Pragmatists rightly point out that these and infinite other social processes of value creation and negotiation are where our ethical deliberations should begin, rather than with a search for fundamental principles of value. Political ecologists examine the social and political interests deeply embedded in this negotiation. In all these cases, how we describe environmental issues, the values invoked in environmental debates, and the variety of meanings brought to these debates are the product of complex social processes that more mainstream environmental ethics has tended to ignore.

The social structures, discourses, and dynamics that are both challenged and, to a certain extent, presupposed by these perspectives represent the relations in which values are established and negotiated. By values, I mean the goods to which we appeal in our deliberations and assessments of an issue, be they environmental, spiritual, social, or something else. Ethics is not reducible to questions of value; whether something is good or of value is a separate question from whether it is morally right to pursue it. Nonetheless, the two questions are related, and it seems correct that what a person regards as a moral obligation depends on what he or she values, on the value or values to which he or she is loyal.[77] For a Christian ethical approach to MTR to respond adequately to the relevant social factors and dynamics, it must examine the values invoked in the debate—identity, authenticity, culture, preservation, and development, to name a few—and the social processes that influence the discursive construction of these values. I have considered how other environmental ethical perspectives attend to these processes of value construction and how they fall short, failing to adequately appreciate the constructedness and relativity of some of their own fundamental values. I now turn to the relational theory of value articulated by H. Richard Niebuhr. More than any of the other perspectives considered, Niebuhr's understanding of value can critique all human values as relational and dynamic without resorting to complete relativism.

A RELATIONAL CONCEPTION OF VALUE

In his essay "The Center of Value," Niebuhr argues that both objective theories of value (which hold that value exists on its own and is inherent in goods themselves) and subjective theories of value (which hold that value exists only in the judgment of an observer, not in goods themselves) invariably move toward a relational conception of value when it comes to ethical questions.[78] For all their efforts to discuss value in the abstract, the former cannot avoid mentioning the being for whom something is good; the latter shift in their ethical considerations from value as a function of feeling to a utilitarianism that takes as its starting point what is good for society. Objective and subjective good, he argues, are ultimately inseparable: what is good or valuable is objectively so, in a way that can be perceived more or less correctly; yet the notion of what is valuable to one being is meaningful only in relation to another being for whom or to whom it is valuable.[79] This is not a matter of simple instrumental value, since every being is both means and end. Whatever is good for me is good not for some finalized, static state of my being; rather, it is good for my becoming, for my progression toward some ultimately unattainable excellence. My progression toward excellence, in turn, is good primarily for other beings and only secondarily for me.[80]

The essence of Niebuhr's theory is that value is objective, yet it is also relational and dynamic: "value is present wherever being confronts being, wherever there is becoming in the midst of plural, interdependent, and interacting existences. It is not a function of being as such but of being in relation to being."[81] Given this web of interconnected relations, he argues, any coherent theory must be based on a center of value (or, more often, multiple centers of value),[82] the being or beings in relation to which whatever is good is good, whether or not this center is explicitly articulated as such. For Niebuhr, all these theories are, in a sense, religious in nature, in that it is impossible to challenge or defend the goodness or value of the center of value itself. Niebuhr argues that most value systems are polytheistic, organized around multiple, sometimes conflicting centers of value.[83] In contrast, the monotheism that Niebuhr advocates proposes a transcendent, resolutely singular center of value with an infinite number of tentative, relative, relational systems of value.[84] Only through confidence in and loyalty to this singular center, which Niebuhr identifies as the principle of being, can moral agents order the manifold value relationships in which they are immersed; only by placing this value center at the center of all their actions, viewing them as responses to this ultimate principle of being, can they find integrity in the multiplicity of actions and claims made on them.[85] According to Niebuhr, the superiority of this understanding of value and the monotheism it supports is principally that only such a theocentric interpretation recognizes

intrinsic value of all creatures, and the rights of all life.[89] A more thorough-going engagement with Niebuhr's relational value theory and monotheism would challenge the apparent self-sufficiency of these values, understanding them instead as involving multiple relations of value and disvalue that are not easily subsumed under such general terms, and relativizing all these centers of value in light of the ultimate center.[90] Whatever community and integrity may mean in the context of environmental ethics, it is clear that they must involve destruction and death as well as sustenance and life. When Rasmussen does consider the inseparability of disvalue from value, near the end of *Earth Community*, the abstraction with which he does so leaves his central ideas of community, integrity, and justice intact. Without a consideration of the specific relationships of value and disvalue in creation, such vague images are unhelpful. Niebuhr's approach engages this multitude of relations, simultaneously relativizing and affirming them by situating them in the context of an ultimate, singular center of value.

This is the particular strength of a relational understanding of value and, more generally, a theocentric ethical approach to MTR. In the interdependent narratives and images of Appalachia—the region's intertextuality—we can see the imbricated relationships that constitute and define values. The insights of the political ecologists are especially useful for tracing out many of these discourses and for calling attention to the operations of power involved in their construction. A theocentric approach, however, resists the temptation to fit this intertextuality into a unifying narrative that would establish moral clarity on a specious foundation, maintaining some stereotypes and dualisms while deconstructing others, as many approaches would do; nor does it document this intertextuality without any real normative assessment, as political ecology tends to do. Rather, because it finds unity not in exclusive categories or narratives but in the universally inclusive center of value, the relational theory of value allows intertextuality to stand, the infinite confrontations of being with being, each one a relationship of value, all of them interpreted and relativized in fidelity to their ultimate center.

Beyond debates about anthropocentric versus nonanthropocentric sources of value, intrinsic versus instrumental value, and objective versus subjective theories, Niebuhr attends to the concrete and contextual relations that are constitutive of value without sacrificing a metaethical foundation for moral discernment. The various dualisms and stereotypes that have pervaded discourses about MTR and Appalachia—narratives about nature and culture, true mountaineers and manipulative outsiders—can be considered and interpreted not on the basis of some predetermined, unstated center of value but as the concrete and contingent interactions of beings and, ultimately, of the principle of being itself, and responded to accordingly. Thus, those who care deeply about both MTR and the fraught and at times oppressive discourses that prevail in discussions of Appalachia and Appalachians would do well to incorporate Niebuhr's relational theory of value and its critical tools into a response to this issue. In the next chapter and those that follow, I describe a response based on Niebuhr's approach in terms of the examination and transformation of imaginations.

3

Relation, Revelation, and Revolution

A Theocentric Approach to Mountaintop Removal

In the previous chapter, I argued that a relational theory of value is better suited to the fraught cultural and historical context of the debate over MTR than other ethical perspectives, specifically in its ability to relativize presuppositions and critique narrow conceptions of value. Yet this critique, however revelatory it may be, offers few resources with which to propose a constructive ethic. My task now is to show how a theocentric ethical approach based primarily on H. Richard Niebuhr's work (including his theory of value) can address and respond to this complex issue and the dynamics surrounding it. In this chapter, I develop this approach as I discern it in the work of Niebuhr himself and that of two Niebuhrians, James Gustafson and Emilie Townes. My theocentric approach, like Niebuhr's own work, is characterized less by concrete prescriptions for specific situations than by an attitude of humility, careful discernment, and critical self-assessment. Nonetheless, this attitude carries real implications for how we interpret and respond to ethical issues, and the work of Gustafson and Townes is particularly useful in exploring these implications. The former elucidates how agents might begin to discern the actions and purposes of God in creation; the latter illustrates the examination and transformation of cultural discourses based on such discernment. My own theocentric approach to MTR incorporates aspects of both. I argue that a theocentric approach calls for a critical examination of the images and narratives used to understand MTR, with particular attention to how the church both challenges and perpetuates images and narratives that have proved destructive and alienating. The goal of this critical self-examination is to recenter the church's imaginations to focus on the action of God the universal valuer, whose purposes relativize all finite interests.

This focus on the church as the relevant moral agent is in keeping with Niebuhr's contention that a believing Christian community aware of its own historical relativity can speak only in the context of its own particular experience. Rather than seeking to justify itself or its superiority over other religious views (and recall that for Niebuhr, all views are in some sense religious[1]), the task of the church is to put forward "a confessional theology which carries on the work of self-criticism and self-knowledge."[2] This is not to say that the church cannot speak meaningfully about the world around it or that it cannot claim to make any true statements about reality; rather, it is to acknowledge that the church's only approach to reality is from its own historically and socially conditioned perspective. As theologian Thomas James puts it, for Niebuhr, "there is no neutral ground on which to stand, and therefore one must simply stand where one finds footing and try to make one's position as intelligible as possible."[3] If the standpoint of the Christian theologian is not the faith of the Christian community, Niebuhr argues, it will be the standpoint of another community with another faith in another god.[4] Thus, the theocentric approach is, from the perspective of the church, the most theologically consistent approach. It is primarily from and to this perspective that I address my approach.[5]

At the same time, the theocentric approach is, if not entirely justifiable, at least commendable to nontheists for nontheistic reasons. By organizing all values around a single transcendent value, the theocentric perspective "destabilizes or fractures the privileged point of view from the inside."[6] Because the center of value is not internal to or wholly understood in terms of the believing community's understanding of itself and the world—because the center is transcendent—the theocentric perspective is capable of radically relativizing its own interpretations in ways unavailable to other worldviews. This is, I believe, theocentrism's signal contribution to ethics in general, including nontheistic ethics: it suggests, in its formal organization around a transcendent center, the possibility of an approach to ethics that radically undermines its own assumptions in order to respond more completely to the multiplicity of interpretations involved in ethical issues. It can be antifoundational without resorting to complete relativism. And it can do this in a way that, as I argued in the previous chapter, seems impossible for approaches that lack a transcendent center. Thus, though I affirm Niebuhr's insistence that there is no neutral perspective, and though much of the theological content of my theocentric approach may be problematic for nontheists, I submit that this approach offers a valuable change of perspective for ethics in general. Those ethics that, like Niebuhr's, seek to destabilize the privileged internal point of view of an ethical community with reference to a transcendent center—and such an endeavor would be consistent with many ethical worldviews, religious and nonreligious[7]—can benefit from the clarity of Niebuhr's moral philosophy. As William Schweiker, another Niebuhrian ethicist, explains, in a "time of many worlds," the imperative of responsibility can both clarify the factors that shape interpretations and provide guidance for meeting new moral challenges.[8] With respect to MTR, the overall shape of my proposed approach may help clarify this complex issue, even if some of the more theologically founded conclusions are unacceptable to nontheists. To put it another way, my approach is both theologically consistent and ethically satisfying; even for those who consider its theological consistency irrelevant, it may prove ethically satisfying nonetheless.

If, as Niebuhr argues, moral action is a response to one's interpretation of events, the adequacy of a response depends on the adequacy of the interpretations involved. I contend that an interpretation of MTR that views the interrelated actions, responses, and imaginations of this issue as somehow reflecting the purposes of God is a theologically consistent and ethically satisfactory interpretation. It is theologically consistent, in that it steadfastly upholds the sovereignty and benevolence of God; it is ethically satisfactory, in that it responds adequately to the various competing value claims (described in previous chapters) by maintaining an orientation toward a transcendent center. The theocentric approach I propose here represents the most (if not the only) satisfactory ethical response to this theocentric interpretation of MTR. Nonetheless, as Niebuhr himself makes clear, a different interpretation, based on a different theology, would evoke a different response, and at the end of this chapter I consider objections to my approach—one ethical and one theological—as well as the question of its anthropological accuracy or usefulness. I maintain that the theocentric approach overcomes these objections and is the most suitable response to such a multidimensional ethical issue.

Niebuhr's Theological Ethics: Three Elements

I have used Niebuhr's theory of value to illustrate how theocentrism allows one to think more critically about some of the complicated questions surrounding MTR compared with many other approaches. Yet Niebuhr is not primarily a value theorist, and theocentrism is not primarily a theory of value. First and foremost, it is Christian ethics. Niebuhr's reflections on value in *Radical Monotheism* and his essay "The Center of Value" can, I believe, be seen as an attempt to describe in philosophical language the fundamental intuitions about value and the church operative in his early works, such as *The Social Sources of Denominationalism* and *The Meaning of Revelation.* Niebuhr begins, in these earlier works, with the historical and social experience of the believing community's faith; he proceeds from there (in his later works) to an analysis of that faith in the more universal terms of valuation.

I note this distinction because, although Niebuhr's theory of value highlights the ability of his ethics to address concerns left unresolved by other ethical perspectives, and although it is indispensable for understanding Niebuhr's thought, his more confessional theological ethic is most useful for articulating a moral response to MTR. More important, in spite of the variety of audiences and languages with which Niebuhr is clearly comfortable, the language of the community of faith is his preferred way of speaking about God.[9] Indeed, as Niebuhr himself acknowledges, there is no other way.[10]

Niebuhr never completed his definitive work on Christian ethics, and his extensive opus represents a wide variety of approaches with appeal to a diverse audience.[11] Nonetheless, it is possible to trace certain key elements of Niebuhr's thought to suggest a coherent ethical approach; this is my goal in this section. First, among Niebuhr's most important contributions to ethical thought is his anthropology—his view of the human person as fundamentally one who responds to the world. This is seen most prominently in *The Responsible Self,* although it is central to much of his work, including his thinking about value in *Radical Monotheism.* Second, attention to this anthropology leads inevitably to a consideration of the being to which the person-as-answerer responds above all else: God. Here, *Radical Monotheism* and *The Meaning of Revelation* are the primary resources for understanding Niebuhr's theology. Third, I turn to the vision of moral action that Niebuhr's anthropology and theology entail. This element of his ethics is most readily seen in *The Meaning of Revelation.*

ANTHROPOLOGY: PERSON-AS-ANSWERER

Niebuhr's theory of value described in the previous chapter reveals the key feature of his anthropology: specifically, the basic relationality that is at the root of all human action. Recall that, for Niebuhr, "value is present wherever being confronts being, wherever there is becoming in the midst of plural, interdependent, and interacting existences. It is not a function of being as such but of being in relation to being."[12] Value is relational; it is a kind of fittingness between beings. In a certain sense, therefore, all value is instrumental: "good" always means "good for."[13] C. David Grant, in his study of Niebuhr's value theory, compares this with a coherence theory of truth. Just as the truth of a statement consists in its coherence with other related statements, rather than any correspondence to an objective reality, in Niebuhr's theory of value, the value of a being is a feature of its relation to other beings.[14] This analogy is somewhat misleading, however, in that coherence with respect to truth presumes noncontradiction: if a set of statements coheres, the statements do not contradict one another. A crucial feature of Niebuhr's theory of value, in contrast, is its recognition that what is valuable for one being in one context may be differently valuable, or even a disvalue, for another being or even for the same being in different contexts. That value exists wherever being confronts being, that it is instrumental in the sense of being valuable for some other being, does not mean that the value relationship is always positive. Being may confront being in a relation of value or disvalue, and each of these occurs in a multiplicity of variations; what is excluded is "non-value."[15] Niebuhr portrays the moral agent as constantly turning from one being to another, buffeted about in a cacophony of inconsistent values.[16]

This relational view of value and human agency is at the root of much of Niebuhr's thought, but it is most fully treated in *The Responsible Self*. Here, he describes the two dominant symbols used throughout history to imagine the human person as moral agent: person-as-maker and person-as-citizen.[17] The former represents a teleological model of action directed toward some end, be it final or proximate, individual or collective. The latter represents a deontological model in which moral decision making is a matter of obedience or disobedience, not of choosing an ultimate end. Niebuhr proposes, as an alternative or complement to both these models, an image of the person-as-answerer.[18] Dialogue and interaction are the fundamental modes of human existence as viewed by psychology, sociology, and biology. Accordingly, "the understanding of ourselves as responsive beings, who in all our actions answer to action upon us in accordance with our interpretation of such action, is a fruitful conception, which brings into view aspects of our self-defining conduct that are obscured when the older images are exclusively employed."[19]

One key to this model of moral agency is the phrase "in accordance with our interpretation of such action." Our moral actions are responses not simply to actions upon us but to our interpretation of those actions as parts of a meaningful whole, and in relationship to the actions and interpretations of others.[20] Niebuhr locates freedom and agency in this moment of interpretation: by choosing or adapting our interpretations, we can modify our responses. Thus, interpretation distinguishes responsive moral action from mere reflexive action. All action has the character of response, according to Niebuhr, but moral action alone is *interpreted* response.[21] That is, only responses characterized by awareness and interpretation of the preceding actions or events can properly be called moral actions, since it is by choosing or changing these interpretations that the agent exercises some control. In place of the teleologist's paradigmatic question "What is my goal?" or the deontologist's "What is the law?" Niebuhr poses the guiding question of an ethic of responsibility, a question that highlights the centrality of interpretation: "What is going on?"[22]

This question is posed on a quotidian level as well as an existential level. In addition to the need to interpret a multiplicity of actions upon oneself as part of a meaningful whole, there is the more basic need to understand one's own existence within that context, to understand "the radical action by which I am."[23] This need can be ignored, or it can be responded to negatively, in distrust, or positively, with confidence and trust in the principle of being at the center of all being. This third response is the response of faith: confidence that the principle of being at the center of all actions and responses is benevolent and trustworthy.[24] This confidence then shapes and transforms the monotheist's responses to all actions upon her.

THEOLOGY: GOD-AS-VALUER

Consideration of the person-as-answerer model of moral action thus inevitably leads to the god to whom one is expected to give an answer. Niebuhr explains his theology in two related ways. The first, taken from *The Responsible Self,* is the one just described: the person-as-answerer seeks, in interpreting the multiplicity of actions upon her, to understand the radical action by which she has come into being. Faith, in this context, is the interpretation of this radical action as trustworthy and benevolent. This interpretation of faith provides the foundation, the meaningful whole, on which all other actions and interpretations can be understood and integrated. Thus, faith in God, whom Niebuhr describes as "the radical action by which I am," is the only source of a coherent identity and agency.

Niebuhr's second explanation of his theology, closely related to the first, appears in *Radical Monotheism*. In the context of his description of relational values, Niebuhr argues that any coherent system of value judgments requires some center of value (discussed in the previous chapter).[25] This center of value is both the organizing principle for all that the self values and the source of the self's own value. My center of value is what I value most, and it determines the extent to which I believe other things to be valuable, including myself. Faith, according to Niebuhr, is the relationship of confidence in and loyalty to this center. It is trust in that which gives value to the self and loyalty to that which the self values.[26] One may have multiple centers of value (polytheism) or turn to one finite center among others (henotheism). However, only monotheism, faith in a single and infinite center of value (identified here as the principle of being), is capable of grounding an integrated and coherent network of value.[27]

Based on Niebuhr's definition of faith, monotheism is trust in and loyalty to the principle of being as the source of all value. This confidence may be expressed in the affirmation that the principle of being, the source of all things, is good for and to the things it creates, in that it gives and conserves their worth. In a sense, this paradigmatic statement of radical monotheism is simply a version of Augustine's "whatever is, is good."[28] In the principle of being, the monotheist finds and commits herself to a benevolent and inclusive object of faith, a center that gives value to all being. In its absoluteness and infinity, this object of faith relativizes all other value relationships.

In these two works of Niebuhr's, God functions (1) as the radical existential action that is the basis for a meaningful interpretation of the various and inconsistent actions upon an agent and (2) as the principle of being at the center of a coherent value system. These are best understood, I believe, as different descriptions of the one role God plays for Niebuhr. In *The Responsible Self*, God is described in the more personal language of human experience of moral action; in *Radical Monotheism*, God is described in the more universal language of value. This is not to say that the two are interchangeable. However similar and mutually supporting they are, Niebuhr's view of value as relational and his view of agency as responsible are not the same thing. Nonetheless, value is a feature of how agents respond to actions, and it names an essential part of the interpretations that shape response. Taken together, these two descriptions of God's role and their corresponding conceptions of faith point to a view of human agency that is thoroughly relational and multidimensional, and they depict a theology that posits God as the only source of coherence and integrity in interpreting and responding to this multidimensionality.

Both these descriptions of monotheism treat faith primarily as a generic, universal human characteristic. The focus is on God's roles in conceptions of moral action and in conceptions of value, respectively, rather than on God as a self that is known in relation to other selves. Elsewhere, however, Niebuhr argues that "one cannot speak of God and gods at all save as valued beings or as values that cannot be apprehended save by a willing, feeling, responding self."[29] His point is that to speak of God abstractly as a source of value or action is inadequate; God can only be spoken of as that which I value or the action to which I respond. Viewed confessionally, from a Christian perspective, the generic faith of *Radical Monotheism* and *The Responsible Self* is secondary—in importance and perhaps chronologically—to the particular revelation of God in Christ. It is not the case that individuals are confronted with the universal need for a unifying object of confidence and fidelity, and God the principle of being fulfills this need; rather, revelation makes this need known even as it fulfills it. It is revelation that exposes how fragmented and conflicted human selves are. In revelation, believers both learn they need a unifying object of faith and find that object.[30]

Niebuhr's understanding of revelation embraces both the particular revelation in Jesus Christ and the entire history in which that event occurs: "The special occasion to which we appeal in the Christian church is called Jesus Christ . . . but from that special occasion we also derive the concepts which make possible the elucidation of all the events in our history. Revelation means this intelligible event which makes all other events intelligible."[31] Niebuhr compares this relationship to a single clear sentence that illuminates an entire obscure book. The particular revelation of Jesus Christ in scripture makes intelligible not only all other parts of that scripture but also all aspects of the history in which the church participates. Thus, revelation involves a dialectic: Christ sheds light on experience, and in this process of illumination, the richness and power of Christ as a revelatory event are made clearer as well.[32] Although revelation, strictly speaking, refers only to this key illuminating event, no aspect of history or experience is excluded from its reach: "nothing God has made is mean or unclean."[33]

Ultimately, this revelation is not simply of an object of faith or a center of value; nor is it simply the principle of being. What is disclosed—or, rather, what discloses itself—in Christian revelation is another self: "Revelation means God, God who discloses himself to us through our history as our knower, our author, our judge and our only savior."[34] What is made known to us is a being that first knows us; what was sought as the universal value is seen, in fact, to be the universal valuer, transforming all finite and partial values. In terms of *The Responsible Self*, the revelation of the radical action behind creation both affirms and relativizes all actions in creation; in terms of *Radical Monotheism*, the revelation of the absolute value and universal valuer behind all value both affirms and relativizes all partial and finite values. This revelation cannot be spoken of impartially, but only from the perspective of faith, of confidence in and loyalty to the knower, author, and valuer of all things.[35]

Niebuhr believes that the fundamental evil in life is inverting this faith, absolutizing what is properly relative. When theology fails to recognize that God, not humanity, is both ultimate value and universal valuer, it inevitably imposes human values on God, finding in God a fulfillment of human needs and interests.[36] This God, he argues, is not the God to whom Jesus prayed, "Thy will be done." In Niebuhr's understanding, when faith in God as the principle of being and universal valuer affirms that all being is good because it is valued, it is decidedly not affirming that all being is good for humanity or for a particular part of humanity. The God that values us does not necessarily value what we value.

Thus, responsibility, monotheism, and theocentrism cohere in Niebuhr's thought as different ways of describing the same ethical model. Through revelation, a person discovers that in and behind all the actions to which she responds (responsibility) are the actions of one principle of being (monotheism). This same revelation transforms her ways of valuing; she places God at their center (theocentrism) and seeks to value as God values and respond to actions as the actions of God. The ethical task then becomes discerning, interpreting, and responding to the values and actions of God. Niebuhr provides some guidance in this regard, which I address next; to some extent, however, this is the task that occupies the rest of this book.

MORAL ACTION: TRANSFORMING EVIL IMAGINATIONS

Whatever is, is good; it is not necessarily right, however. Whereas all beings are valued, and valuable, in relation to God, the relationships between beings can be ones of disvalue, often great disvalue.[37] What guidance, then, do the images of God-as-valuer and person-as-answerer provide in distinguishing right relationships and actions from wrong ones?

Some general features of right moral action have already been noted.

Right moral action is responsible, in the sense that an agent interprets all proximate actions as expressing the actions of God and responds to that ultimate source. Similarly, because God is the center of value, a Niebuhrian monotheistic system of value organizes all value relationships around God, and because God is the universal valuer, such a system values all finite beings, albeit in a relative way. Based on these basic features and on Niebuhr's lecture notes, Grant identifies four principles for moral decision making: "They are: (1) 'Forget about your own value. You are beloved. You are saved.' (2) 'Serve the value which is in greatest need.' (3) 'Serve the values at hand—those near to you.' But perhaps the most significant of those principles is the last: (4) 'There is no value choice we make which must not be made in reliance on forgiveness because it involves sacrifice of the good.'"[38] Thus, according to Grant, radical monotheism provides no positive criteria for decision making; rather, it shapes a fundamental attitude characterized by humility and an awareness of the finiteness of all human judgments of value. In this sense, for Grant, Niebuhr's ethic is exclusively an "agent ethic" rather than an "act ethic."

It is possible, however, to discern a more positive ethical approach in Niebuhr's work. Although he steadfastly eschews concrete ethical norms (a choice that is consistent with his understanding of the contextual, historical character of all human action), Niebuhr describes how his radically monotheistic faith transforms the way believers respond to their world.

As noted earlier, the key moment of freedom and agency in human moral action is the moment of interpretation; it is here that the agent exercises control over her responses to actions upon her. Niebuhr argues that this interpretation takes place with the aid of images and symbols.[39] Images and imagination are not, as many would argue, irrational, a function of the heart rather than the mind; according to Niebuhr, the imagination works together with reason to interpret the data of the world around us.[40] Prior to revelation, our images tend to be inadequate at best and destructive at worst: they place ourselves at the center of our world, alienating us from others and making us unable to integrate and understand the many divergent actions and events of our histories. The fruits of these images—conflict, destruction of self, a world of confused agencies and divided loyalties—reveal them to be evil.[41] Revelation—the moment in the community's history that illumines the rest of that history (for Christians, the event of Jesus Christ)—displaces these destructive images and renders not only the past but also the present and the future intelligible as one whole. The revelation of God as the self that values all other selves displaces our own selves from the center of all our images, revealing clearly both the unity of all being and the breaches of that unity caused by our selfish, destructive imaginations.[42] Revelation creates revolution in all our notions of value, goodness, and power.[43]

Given this understanding of the transformative power of revelation, the goal of Christian ethics must be to maintain this decentering, critical challenge to all destructive, self-centered imaginations. As the community of this revelation, the church must undertake a careful, critical appraisal of its own participation in the evil imaginations of the world and seek to transform them in light of the God who values all being.[44] The church has no cause to boast; believers may only confess: "we were blind in our distrust of being, now we begin to see; we were aliens and alienated in a strange, empty world, now we begin sometimes to feel at home; we were in love with ourselves and all our little cities, now we are falling in love, we think, with being itself, with the city of God, the universal community of which God is the source and governor."[45] Sustaining this transformation is the central imperative of a Niebuhrian ethic, and it is the foundation of the theocentric approach I propose.

Toward a Theocentric Ethical Approach: Two Niebuhrians

This move toward a positive theocentric approach to ethics remains abstract, partly because of Niebuhr's insistence that moral decisions are irreducibly contextual and historical. Nonetheless, this approach can be clarified by considering the work of two prominent Niebuhrian ethicists: James Gustafson and Emilie Townes.

DISCERNMENT AND HUMILITY: JAMES GUSTAFSON

James Gustafson is the leading interpreter and exponent of Niebuhr's thought. In his definitive work *Ethics from a Theocentric Perspective*, Gustafson takes up Niebuhr's criticism, found most prominently in *The Meaning of Revelation*, that most mainstream theology, even when it purports to be theocentric, imposes human values and interests on its understanding of God, making God beholden to human needs. Like Niebuhr, he sees this as the perpetual temptation of religion.[46] In terms reminiscent of his mentor's responsibility ethic, Gustafson concludes that the guiding question for a theocentric ethic is "What is God enabling and requiring us to be and to do?"[47] The general response to this question, and the moral imperative of theocentric ethics, is to "relate to all things in a manner appropriate to their relation to God."[48] In any particular situation, this requires attending to the divine ordering of creation as it can be perceived, however faintly, in the patterns and processes of interdependence between human life and the wider world.[49] This, in turn, leads Gustafson to a more inclusive understanding of the "relevant wholes," the individual or collective units that must be considered when evaluating courses of action, and to a broad but nonetheless careful incorporation of insights from a wide array of disciplines such as the social sciences, economics, and (most prominently) the natural sciences. He applies this approach to four specific moral areas: marriage and the family, suicide, population and nutrition, and the allocation of medical research funding. In each case, he examines the relevant wholes and the relationships and dynamics involved and then draws on these examinations to discern what God is enabling and requiring with respect to that moral area.

Gustafson's work helps clarify how the transformation of imaginations and the discernment of God's purposes might be undertaken and sustained. The revolution that replaces finite selves with God's purposes at the center of our imaginations surely requires that the morally relevant wholes be conceived more inclusively. Meanwhile, both the central question of "What is going on?" and the belief that the revelatory event of Jesus Christ illuminates all of history invite the church to use any means available to understand more fully the world around it, as exemplified by Gustafson's careful interdisciplinary work.

These common themes notwithstanding, Gustafson (at least in *Ethics from a Theocentric Perspective*) is more optimistic about the legibility of God's purposes in the patterns and processes of creation and less attentive to the particular history of the church, including revelation in Jesus Christ, than Niebuhr is. Gustafson himself identifies the main reason for this as his and Niebuhr's divergent theologies: "He had more confidence in the agency model of God than I have . . . I believe we can appropriately say only that we have capacities to respond to persons and events . . . and that through those actions we respond to the divine governance."[50] It is this suspicion of a conception of God as agent, as well as his affirmation of the Calvinist view that "the divine law is present in the natural ordering of things, and thus action is to be in accordance with the natural law," that leads Gustafson to rely much more heavily on the natural sciences and other disciplines, and less on the particular perspective of the Christian community, than his mentor.[51] His confidence that God's purposes can be seen, however partially and faintly, in the created order allows Gustafson to suggest clearer ethical guidelines for the specific issues he considers than those provided by Niebuhr.

In this respect, my proposal adheres more closely to Niebuhr's view, primarily because I take seriously his contention that the church ultimately speaks only from and to its own particular perspective. In addition, I believe that this perspective offers certain insights about God's purposes that are unavailable from other perspectives. This also means that, like Niebuhr, I am more reluctant to propose clear guidelines about the issue at hand than Gustafson (who is nevertheless quite cautious), although I do provide some guidelines in the final chapter that I believe issue from my approach. Nonetheless, Gustafson's inclusive and interdisciplinary work contributes much to my own application of Niebuhr's thought to MTR.

In his much smaller, less widely known book *A Sense of the Divine: The Natural Environment from a Theocentric Perspective*, Gustafson offers more modest suggestions for thinking ethically about environmental issues. Four related elements—theocentrism, a relational or multidimensional theory of value, recognition of humans' place as participants, and moral ambiguity—cohere in his approach to the environment. The first two have already been discussed; the third, participation, points to humanity's role as participants in an interactive network of value, rather than as objective observers. Because of this interaction, consideration of value must necessarily ask whose value and whose good is being pursued, and at whose expense. The fourth element, moral ambiguity, argues that because of this relationality and multidimensionality, moral choices often involve sacrificing one value for another (recall that this was one of the key features of the Niebuhrian attitude outlined by Grant). However appealing the idea of harmony may be, whatever equilibrium exists in nature is always dynamic and imperfect, involving death and disorder as much as life and order.[52] For environmental ethics, this means that pursuing one value frequently involves accepting the sacrifice of other values. The moral attitude that results from these four elements affirms that God is the orderer of all creation, but God's order has no clear or apparent telos, at least from a human perspective. As participants, we humans are both accountable and ultimately dependent, called to intervene in some cases, but always with a humble recognition of the ambiguity of all our actions.[53]

Gustafson offers fewer moral guidelines in *A Sense of the Divine* than in *Ethics from a Theocentric Perspective*. The multidimensionality of value and the ambiguity of moral decision making call for sensitivity to the various contexts in which decisions are made. Indeed, after considering a variety of moral discourses relevant to environmental issues, he concludes that the theocentric perspective recognizes the difficulty of reconciling them, and he attributes this to the ultimate limitations of human finitude.[54] Nonetheless, in his attention to these discourses, certain features of his own position are made clear: an epistemic humility that sees the multivalence of any potential outcome, an awareness that the way one defines an issue reveals certain selection criteria and priorities, and the need to proclaim certain attitudes and exclude others while simultaneously attending to the complexity and nuance of any given debate.

Among the moral discourses Gustafson considers, the one most relevant to this study is what he describes as a narrative approach, which seeks prophetically to raise consciousness and decry certain evils, yet with a sense of ambiguity.[55] He points out that narratives frequently carry strong moral valence and can evoke clear emotional connections between humans and nonhumans. These emotional connections and the narratives that foster them seem to be part of the "sense of the divine" of the title, the intuitive experience of nature that is at the foundation of a theocentric moral stance. At the same time, he is cognizant of the complexities within any one narrative and the conflicts between narratives, and he acknowledges that narratives do not resolve social or policy decisions. Narratives shape attitudes and actions, but they are also shaped by them, in a variety of ways.[56]

The applicability of these observations to MTR is clear. The moral weight of the stories told—stories of homes and cemeteries destroyed, families sickened, and communities torn apart—is unmistakable. Yet equally undeniable is the complex interrelation of these narratives and others, as well as the fact that the stories alone, for all their compelling emotion, are unable to resolve the complicated moral questions they raise.

Like prophecy, narrative has a significant impact on how people are affected by and respond to moral issues; like prophecy, it cannot ultimately resolve those issues.[57] In the context of *A Sense of the Divine* and of Gustafson's work more generally, it is clear that, for him, narrative represents a valuable part of human experience of the world, but one that must be duly chastened by dependence and accountability. If this view of narrative is broadened to include all the interpretive resources humans use to understand their world—what Niebuhr describes as imaginations—then this aspect of Gustafson's ethic has much in common with my theocentric approach. In addition to incorporating the various disciplines considered in *Ethics from a Theocentric Perspective*, a theocentric approach to environmental ethical issues draws on narratives and other "cultural penetrations of nature" as part of the truth of the human encounter with the world—an encounter that creates the possibility of a sense of the divine.[58] Yet this approach also examines how these narratives undermine accountability before and dependence on God.

CHALLENGING EVIL IMAGINATIONS: EMILIE TOWNES

In *Womanist Ethics and the Cultural Production of Evil*, ethicist Emilie Townes applies the Niebuhrian examination of imaginations to the cultural construction of race and gender. Believing that "to simplify the complex is to neuter richness and defame the marvelously complex gift of life found pulsing ubiquitously around us," Townes engages Niebuhr's paradigmatic question, "What is going on?" by interrogating several racist and sexist images—the Black Mammy, Sapphire, the Tragic Mulatta, the Welfare Queen, and Topsy—to show how evil is produced and sustained in these images and how it can be challenged and undermined by them as well.[59]

Echoing Niebuhr's contention that the interpretation that characterizes moral action utilizes the imagination and that responsible action requires the examination and displacement of destructive imaginations, Townes argues that racist and sexist images are part of a "fantastic hegemonic imagination" that distorts and fractures our interpretation of the world. Based on the work of Antonio Gramsci, she argues that hegemony is sustained by an exclusive understanding of history that forestalls wholeness by insisting on a single, authoritative (and inevitably oppressive) interpretation of the past. Counterhegemony and its tool countermemory, according to Townes, can challenge hegemony by disrupting ignorance and invisibility and calling attention to the particularities overlooked in hegemonic histories.[60] Countermemory recovers the lived experience behind stereotypical images like the ones Townes examines in order to reconstitute shared histories, offering more integrated and integrating self-understandings.

The point is not that countermemory is more accurate than history by any objective standard; indeed, claims of accuracy and objectivity are the means by which hegemony seeks to maintain its dominance. Yet because it operates from the perspective of those oppressed by hegemonic imaginations, countermemory offers a view of the "interior worlds of structural evil" to present a fuller, more nuanced, and more complex picture of the world in which moral action operates, thereby beginning to dismantle destructive imaginations. Countermemory and counterhegemony, by their very nature, are multifaceted and complex; they interweave the multiple and at times conflicting relationships of persons' lived realities. Counterhegemony is opposed to hegemony not as one truth to another but as the multiple imbricated truths that challenge hegemony's pretense of exclusivity.

In the Gramscian conception, hegemony operates through the establishment of one coherent dominant worldview, such that the oppressed effectively consent to their oppression. Neither hegemony nor counterhegemony is the exclusive property of one group or another, as Townes notes.[61] Accordingly, the destructive images she explores also carry the seeds of counterhegemony and countermemory, unsettling and disrupting the very hegemonic notions they were originally mobilized to establish.[62] As Townes explores each image in turn, she brings out these threads of resistance. For example, with respect to the Mammy image, she points to the double meanings in the folk figure of Aunt Jemima, simultaneously supporting white superiority while subtly undermining it.[63]

Of particular interest is Townes's analysis of the image of Sapphire, a negative stereotype of black women as "malicious, vicious, bitchy, loud, bawdy, domineering, and emasculating."[64] Townes argues that these are precisely the characteristics necessary to disrupt the discourses of "uninterrogated coloredness" that underlie the cultural production of race and racism. "Sapphire," she argues, "is based on the oldest negative stereotype of women: inherently and inescapably evil. Perhaps in the case of racism, only a stereotype or an image that is based on evil can help destabilize and deconstruct a structural evil such as racism."[65] The characteristics of Sapphire are capable of confronting the "juggernaut" of racism in its manifold permutations and penetrations of culture. In particular, from the perspective of the collective experience of black women, the figure of Sapphire calls our attention to history and tradition, especially those histories and traditions that have been excluded from the dominant discourse. Sapphire asks, "How can an authentic ethic of justice be separated from where we have been and who we have been to one another?"[66] In the figure of Sapphire, as in all of Townes's images of black womanhood, we see that in destructive or inadequate imaginations, hegemony and counterhegemony are thoroughly intertwined. Townes's careful examination seeks out the latter in order to challenge and weaken the former, to begin the long process of dismantling cultural evil.

This is a thoroughly theocentric project: seeking the transformative movement of the divine in the most oppressive and destructive cultural imaginations and replacing the finite and divisive human interests at their center with the inclusive, universal, unifying valuer of all being. Although the terms counterhegemony and countermemory are not Niebuhr's, they help clarify the reinterpretive work of theocentrism. To counter evil imaginations that place selfish human interests at their center, thereby alienating beings from one another and destroying the unity of creation, he offers theocentric imaginations that seek God's action and inclusive purposes in the multivocality of creation; in this sense (though not in a more political, Gramscian sense), the theocentric approach is counterhegemonic. The substance of Townes's work in *Womanist Ethics* is strikingly similar to Niebuhr's call for self-examination, repentance, and renewal. He exhorts the church to examine its participation in the evil imaginations of the world and to transform those imaginations around the single, universal value center that is God.[67] Townes does not address the church specifically, but her critical examination of discourses of race and gender parallels his work. Sapphire's question, "How can an authentic ethic of justice be separated from where we have been and who we have been to one another?" can be read as one variant of Niebuhr's paradigmatic "What is going on?"

A Theocentric Approach to Mountaintop Removal

Townes's focus on racial and gender discourses may seem, at first glance, to have little to do with MTR. Yet what she calls the fantastic hegemonic imagination, the set of destructive cultural images and stereotypes that support inadequate and alienating worldviews, is at work in the divisive and culturally fraught discourses surrounding MTR. As already noted, the debate over MTR perpetuates and updates narratives that have dominated and manipulated understandings of Appalachia for more than a century. Hegemony and counterhegemony are interwoven and interact in narratives about Appalachian identity, about power in the Appalachian context, and about destruction and reclamation. Sapphire's question could easily be asked of Appalachia. Moreover, Townes's (and Niebuhr's and Gustafson's) recognition of the need for complexity and nuance in examining how our values and actions are shaped in interaction with these imaginations is essential for approaching the intertextuality of Appalachia.

My theocentric approach is therefore similar to Townes's approach, although it has a different (and arguably more Niebuhrian) starting point: the church. As noted at the beginning of this chapter, Niebuhr insists that an awareness of historical relativity leads the church to understand its task as one of confessional self-criticism from its own particular perspective.[68] I follow Niebuhr's example in *The Meaning of Revelation* in this regard, speaking from and to the perspective of the church; however, this does not mean that the approach I propose is of value only for that particular community.

For Niebuhr, revelation comes in and through history; this requires that the church attend carefully to its participation in and perpetuation of histories. This calls for scrutiny of the church's complicity in destructive imaginations so that those imaginations can be reorganized around their proper universal and unifying center.[69] I propose, therefore, a theocentric approach to MTR that examines key narratives surrounding it—I focus on discourses of power and powerlessness, insiders and outsiders, and destruction and reclamation—and the church's use of these narratives. I agree with Townes's belief that counterhegemony and countermemory are necessary tools for challenging and reconstituting the destructive imaginations Niebuhr describes. The goal of this approach, then, is to cultivate these tools to provide a fuller picture of "what is going on" and to seek the movement of the divine in a complex and intertextual reality. As we have seen, hegemony and counterhegemony are intertwined in cultural imaginations, and the church participates in both. My intention in discussing each set of ideas is to suggest how the church colludes in as well as challenges the manipulation and construction of the image of Appalachia. My conviction is that this fuller awareness is the first step toward the transformation of values that Niebuhr describes, an attempt by the church "to see the reflection of itself in the eyes of God."[70] In this attempt, like Gustafson, I seek to clarify God's action in the world through attention to the insights of a variety of disciplines (although I draw more on the social sciences than the natural sciences), and I argue for a more inclusive conception of relevant wholes.

The primary goal of this approach is not to offer concrete prescriptions for action. Rather, the goal is to destabilize self-centered, destructive, or alienating imaginations in order to consider how the church might recenter its interpretations around the God whose values are not our own. This is not to say, as Gustafson does, that human well-being is not one of God's values; from my perspective, as from Niebuhr's, God's benevolence toward creation, including humans, is a matter of faith: God "ministers indeed to all our good but all our good is other than we thought."[71] The point is that

imaginations centered on the God who values all being are radically different from limited, self-centered imaginations.

To summarize: a Christian response to MTR, developed along Niebuhrian theocentric lines, engages in a humble and critical self-assessment
of the church's complicity in and challenge to the divisive, destructive, or
inadequate imaginations that continue to shape society's understanding of
MTR—that is, the church's participation in both hegemonic and counterhegemonic imaginations. This critical examination is not undertaken for its
own sake, as an exercise in self-abasement; rather, its purpose is the transformation of our interpretations of the world around a new focal point. The
goal is to clarify the church's response to the question "what is going on?"
by seeking those images and interpretations whose central values are the
purposes of God, rather than human values and interests, in the intertwin-
ing actions and relations to which the church responds.

This theocentric approach to MTR is, I believe, both ethically supe-
rior and theologically consistent. In the previous chapter, I showed that
Niebuhr's understanding of value enables his ethical approach to be more
attentive to the complex issues involved in MTR than other environmental
ethics can be. Here, I have argued that the theology at the heart of Niebuhr's
ethic (and my application of it) is thorough and consistent in its insistence
on the sovereignty of God. Nonetheless, one may object to this approach
on both ethical and theological grounds; that is, one may argue that it is
insufficient to support a strong ethical challenge to apparent injustices, or
one may argue that the approach's theology is inadequately Christ-centered.
I address each of these objections in turn, and then I consider the anthropological adequacy of Niebuhr's model: Is it an accurate, or at least useful,
portrayal of human moral agency?

AN ETHICAL OBJECTION

On its surface, this response may appear quietistic and conciliatory: seeing
the purposes of God in a practice as destructive as MTR seems to leave no
possibility for a prophetic challenge of this (or any other) practice as sinful.
On one level, this is precisely the point: the prophetic voice, no less than the
voice of hegemony, is tempted to place finite selves and human values at the
center of its moral critique. Niebuhr argues that "the great source evil in life
is the absolutizing of the relative, which in Christianity takes the form of
substituting religion, revelation, church or Christian morality for God."[72] If
a strong prophetic challenge to MTR involves this kind of absolutizing of

the relative, it is theologically suspect. Likewise, if the challenge relies on incomplete histories or alienating imaginations, it is ethically inadequate and should be questioned and scrutinized.

Yet what this discussion neglects is the genuinely radical nature of this ethic. As Douglas Ottati notes in the introduction to *The Meaning of Revelation,* H. Richard Niebuhr, like his brother Reinhold, begins with the desire to "distinguish an integral Christian ethic from the dominant interests of modern western culture."[73] In distinction from Reinhold, however, H. Richard locates this difference at the level of conflicting fundamental faiths. Mainstream society differs from the community shaped by Christian commitments because they orient their values and interpretations of the world around entirely different centers. The events and actions that each community interprets and to which they respond are not essentially different, but the characters of their responses are divergent because they arise from drastically different interpretive foundations. Niebuhr's monotheism is therefore radical in the most basic sense of the word, challenging the very roots of how we interpret and engage moral decisions. His position and the ethic it supports are indeed prophetic; they are made all the more so because, in this case, the prophetic voice is leveled against the church and ostensibly Christian morality rather than against "the world," as if the former were innocent of the sins of the latter. Niebuhr recognizes, as Gustafson and Townes do, that a thoroughgoing monotheism challenges, relativizes, and transforms moral action at all levels.

A THEOLOGICAL OBJECTION

A second challenge leveled at Niebuhr's theology is that his monotheism neglects the person of Jesus Christ in favor of a radically transcendent God.[74] A more Christocentric theology would, it is argued, result in a different (likely a more directly political) ethic. Again, Niebuhr is self-consciously aware of the temptation to substitute any source of value—even Jesus Christ—for God.[75] Part of the justification for avoiding this substitution is that it misunderstands the significance of Jesus. This focus on the historical person of Jesus inevitably leads to a focus on the church as the revelation and incarnation of Jesus, and once again, the self and its interests displace God as the center of value.[76] Although he is mindful of this risk, Niebuhr is no less insistent that Jesus is the decisive moment of God's revelation.[77] In the life of Jesus, in "the fullness of time," all that came before and all that comes after are made intelligible as part of the divine purpose. The Christ event is the lens through which believers organize their interpretations of events and actions upon them. Jesus makes it possible for believers to see in these divergent and multifaceted events and actions the purposes of a value center that is both powerful and benevolent.[78] Thus, while Niebuhr is critical of an exclusive focus on the life and teachings of the historical Jesus as a source for ethics, he is clear that it is not monotheism in the abstract but belief in (that is, confidence in and commitment to) the God of Jesus Christ that radically transforms all values.

AN ANTHROPOLOGICAL QUESTION

Finally, one might ask whether moral agents really operate as Niebuhr believes they do. Is his responsibility model more accurate than the deontological or teleological model? To answer this question definitively would require empirical exploration beyond the scope of this study. Yet Niebuhr believed that the insights of various sciences—biology, psychology, and sociology—supported his view of human agency. Some students of Niebuhr, from Gustafson and Townes to Stanley Hauerwas and Larry Rasmussen, have found his responsibility model of moral action convincing, despite their other differences with him.[79] This view of human agency has found clear resonance with many thinkers.

More important, though, is Niebuhr's insistence that each of the different models commends itself based not on some comprehensive accuracy but on its ability to make certain features of life more understandable. Responsibility, he argues, is not an absolutely new and complete conception of human moral life. "Actuality always extends beyond the patterns of ideas into which we want to force it. But the approach to our moral existence as selves, and to our existence as Christians in particular, with the aid of this idea makes some aspects of our life as agents intelligible in a way that the teleology and deontology of traditional thought cannot do."[80] If responsibility as a model of agency sheds light on features of human life that had previously been neglected—that is, if it is useful—then it is, in that limited sense, accurate. For example, Niebuhr argues that suffering, as the intrusion of elements that thwart our purposes, cannot be understood adequately by teleology or deontology, whereas his responsibility ethic is able to comprehend it more fully as something that an agent interprets and to which she responds.[81] It is, in this respect, useful: it sheds light on the experience of suffering.

In a similar way, just as Niebuhr's value theory is particularly useful in making moral sense of the divergent narratives of value surrounding MTR,

I believe that his model of moral agency illuminates features of the church's responses to those narratives in a way that other approaches have been unable to do. In other words, in addition to being theologically consistent, in that it maintains a steadfast focus on the purposes of God as revealed in the person of Jesus Christ, a theocentric approach is ethically useful, in that it begins to make moral sense of the complexities of the issue at hand and offers resources for ethical responses.

In this chapter, I have outlined a theocentric Christian ethical approach to MTR by tracing key features of Niebuhr's theocentrism—specifically, his anthropology and theology and the ethics they entail. I have argued that a theocentric approach would be characterized by a humble and clear-eyed assessment of the imaginations the church relies on to interpret the events and issues it responds to and how those imaginations might be destructive or inadequate. The work of James Gustafson and Emilie Townes provides further clarification of how this attitude might be applied to specific questions. Specifically, Gustafson's work illustrates the broad exploration needed to discern God's purposes in creation, as well as the humility with which this discernment is undertaken, and Townes offers examples of the examination and rehabilitation of imaginations, seeking counterhegemony in the tools of hegemony.

My approach consists in exploring in detail certain key ideas and narratives that have given the debate over MTR its destructive polarizing force, as well as the different ways the church has engaged these ideas, in the hope of reconstituting the church's images to reflect the inclusive purposes of the God who values all being. I do not imagine that this approach will necessarily be more effective in practical terms than other approaches, nor that it will lead to reconciliation between divergent parties. Rather, this approach is an ethically useful response to MTR, in that it is better able to appreciate and critique the intersecting narratives that shape this issue. Moreover, this approach is a theologically consistent response, in that it alone resists absolutizing finite human values and identities, focusing instead on inclusive divine purposes. If moral action has the fundamental character of response, a theocentric approach responds to the events and actions surrounding MTR as reflecting God's purposes in creation, however inscrutable those purposes may be. It does so by examining the imaginations that interpret situations and shape moral action in order to replace destructive imaginations with God-centered ones and to replace the human interests at the center of those imaginations with inclusive divine purposes. In the following chapter, I apply this project of critical examination to some of the key imaginations of MTR: specifically, notions of power and powerlessness, insider and outsider, and destruction and reclamation.

An active mountaintop removal site. A dragline can be seen in the center of the photograph. (Photo courtesy Vivian Stockman/www.ohvec.org. Flyover courtesy Southwings.org.)

The active mine site on Kayford Mountain, near the home of Larry Gibson. (Photo courtesy Vivian Stockman/www.ohvec.org. Flyover courtesy Southwings.org.)

One example of reclamation. Note the sparse trees and the rolling grassland. (Photo courtesy Vivian Stockman/www.ohvec.org.)

Another example of reclamation, the infamous Twisted Gun Golf Course. (Photo courtesy Vivian Stockman/www.ohvec.org. Flyover courtesy Southwings.org.)

Stan Holmes leads a "Blessing of the Mountains" service at the entrance to an active mine site. After an initial confrontation, many of the participants engaged in a frank and open discussion with miners. (Photo courtesy Vivian Stockman/ www.ohvec.org.)

Stover Cemetery on Kayford Mountain, surrounded by an active mine site. The cemetery is the small dark patch near the center of the photograph. (Photo courtesy Vivian Stockman/www.ohvec.org. Flyover courtesy Southwings.org.)

4

The Meanings of
the Mountains

Discourses of Power, Identity, and Destruction in the Mountaintop Removal Debate

With respect to MTR, Niebuhr's central question, "What is going on?" goes beyond the standard dilemma of jobs versus environment that is so often invoked in this and other environmental debates. It points to more fundamental questions about the meaning of Appalachia itself. The idea of Appalachia as a discrete region with its own peculiar ecology, people, and customs is the result of a long series of interrelated discourses. In this chapter, I examine those discourses and focus on three pairs of concepts—power and powerlessness, insiders and outsiders, and destruction and reclamation—that are particularly relevant to MTR and particularly rich territory for an analysis from a theocentric perspective. First, however, I discuss how these ideas have functioned in descriptions of the region and debates about MTR and explain why their analysis is important. Then I turn to the concepts themselves.

The Intertextuality of Appalachia

Discussions of Appalachia and its peculiar struggles tend to follow familiar patterns. Although the region may be variously characterized as a place of pristine natural beauty or the source of valuable domestic energy, it is nearly always described as isolated or set apart in some way. Appalachians are seen as either marginalized victims of a global economy or environmentally ignorant hillbillies; in any case, they are portrayed as simple folk having some special tie to the land, as either farmers on it or miners beneath it. The enemy is always an outsider: manipulative environmentalists from

California or Florida, controlling bureaucrats in Washington, DC, or greedy corporations in Virginia or Missouri.

These characterizations—often superficially contradictory—draw on a long tradition of representation and manipulation and are rooted in beliefs about the identity of the United States itself. Before the Civil War, the "discovery" of a region that was part of the nation yet apparently separate and different from it presented a problem for the idea of a unified national identity. This paradoxical otherness has continued to characterize Appalachia in the eyes of the larger society. The past and present negotiation of this identity is addressed in greater detail later; the point here is to note that ideas about Appalachian identity have always been constructed along different lines to serve a variety of political interests, yet the characterization of the region as a problematic other has remained constant.

Anthropologist and Appalachian scholar Allen Batteau argues that these kinds of discourses are false not in their substance but in their social function. The portrayal of Appalachia as an alien region and a client population, however well intentioned, misrepresents life in the Southern Mountains by treating the region's otherness as a fact to be explained rather than recognizing it as an idea constructed and manipulated for various political ends.[1]

Geographer Stephen Hanna contends that even astute political treatments like Batteau's do not go far enough. While they challenge the stereotypes and assumptions that have dominated understandings of Appalachia for more than a century, they implicitly allow the dualisms that founded those assumptions—dualisms such as insider-outsider and hegemony-resistance—to remain. According to Hanna, authors like Batteau and fellow Appalachian scholar Henry D. Shapiro present the manipulations of Appalachianism as being imposed by outsiders and as misrepresenting the "real" Appalachia known and accessible to "natives."[2] This implicit dualism undermines the authors' own goals. Identifying a hegemonic narrative and treating it as monolithic, for example, implies that there is only one pervasive and defining discourse about Appalachia; focusing on outsiders' perceptions of Appalachia undervalues the agency of those who live in the region and the power of their perceptions.

Scholars must therefore subvert these dualisms themselves, as Hanna argues, and challenge the idea of a "real" Appalachia, positing instead a system of related and mutually dependent understandings and identities. Hanna calls this network of representations the intertextuality of Appalachia, a term I have adopted for this book.[3] The multivocal texts of Appa-

and responds to the fundamental integrity and divinity of all creation; other interpretations only contribute to the divisiveness and confusion of human agency.[86] As I argue here and in later chapters, the difficult case of MTR bears this out: a theocentric approach based on Niebuhr can make moral and theological sense of competing value relations and the discourses that shape them in a way that other perspectives have been unable to do.

This theocentric approach in general and its implications for a Christian ethical response to MTR are developed in detail in the following chapters. Here, however, I want to focus on the relational conception of value, arguing that it is uniquely suited to the multidimensionality and complexity of cultural narratives surrounding MTR and Appalachia in general. The relational theory of value can grasp and appreciate the intertextuality of the region, with narratives building on other narratives without imposing preconceived ideas of a "real" Appalachia. Categories of insiders versus outsiders, environment versus jobs, true Appalachians versus true Americans can be seen as interacting attempts to define and defend value in the context of myriad concrete relationships that are not simply asserted as true or false, authentic or distorted, oppressive or liberatory.[87] In this way, this approach relativizes all assumptions, all preconceptions, and all attempts to assert value not centered on the principle of being. In contrast, each of the first four environmental ethical perspectives discussed earlier, while attentive to the social construction and negotiation of value, is limited in its ability to question the value-laden assumptions on which it is founded: ecofeminism, liberation theology, and environmental justice fit MTR and other issues into a predetermined framework of injustice and justice or oppression and liberation, while pragmatism intentionally avoids these metaethical concerns through a narrowly practical focus. In doing so, these approaches fail to see that categories coexist in mutual interdependence. The fifth approach, political ecology, turns a critical eye on precisely these issues of socially constructed values, yet without coherent recourse to its own ethical norms.

Even the compelling work of Niebuhrian environmental ethicist Larry Rasmussen fails to apply Niebuhr's value theory consistently and thus insufficiently destabilizes dualistic stereotypes. His *Earth Community, Earth Ethics* engages Niebuhr's central question of "what is going on?" and its variants to discern the actions of God in environmental crises, and it advocates a Niebuhrian responsibility to God for "all that participates in being."[88] Yet the central value that orients this responsibility, "sustainable community," is left uninterrogated, as are other key concepts such as the integrity of creation, the lachia—narratives, identities, stereotypes, images in popular media—are constructed on top of one another, rather than deriving straightforwardly from some fundamentally real experience of the region. Hegemonic and resistant narratives, and identities of insider and outsider, depend on and produce each other and therefore always coexist: "the existence of resistance within hegemony, and hegemony within resistance, ensures the continued reproduction of alternative representations and meanings."[4] Identity exists not within but rather across and around these categories.

If, as discussed in the previous chapter, moral action is informed by the imaginations we use to interpret our world, then these discourses shape our imaginations of Appalachia and MTR and our actions in this regard. Moreover, insofar as these imaginations place finite selves and human interests at their center rather than divine purposes, they are, from a theocentric perspective, inadequate and destructive, leading to alienation from self and others rather than integrity and responsible action.[5] A theocentric Christian approach thus requires careful analysis of these discourses, their hegemonic and counterhegemonic implications, and the church's participation in and perpetuation of them. This is undertaken not only to show that these politically constructed discourses are incomplete but also (and more importantly, from a theological-ethical perspective) to search for more theocentric imaginations.

Given the rich intertextuality of Appalachian discourse, such an analysis could focus on a number of concepts—nature and culture, oppression and resistance, progress and development, race and gender. The three pairs of concepts I have chosen are especially prominent in (or just beneath the surface of) debates about MTR and are particularly fruitful for an exploration along Niebuhrian lines: power and powerlessness, insiders and outsiders, and destruction and reclamation. For each of these binaries, I describe how the pair has been politically constructed in discourse about Appalachia, indicate how it has been mobilized in the debate over MTR (both in general and by the church), and suggest some of the inadequacies of this mobilization. In considering the church's use of these concepts, I draw on a variety of sources: the Roman Catholic pastoral letters already discussed, resolutions from national church bodies (specifically, the Evangelical Lutheran Church in America, the Episcopal Church, the United Methodist Church, the Presbyterian Church [USA], the Unitarian Universalist Association, and the Religious Society of Friends, as well as the West Virginia Council of Churches), articles in Christian publications, and informal interviews (my own and others') with prominent figures in the discussion. I do not assume that these sources are authoritative or illustrative of all (or most) Christians' attitudes toward MTR, or that there is agreement among or even within the various traditions about these questions. My purpose is to show that the discourses I describe have influenced the way some Christians talk (and presumably think) about MTR in a variety of settings and to argue that, to the extent these discourses are inadequate, Christians have good reason to reevaluate their use of them.

After examining each of these pairs, I argue that they are all part of the construction of the meaning of place in Appalachia, but that this understanding of place, like the other concepts discussed, is not as straightforward as many have assumed. The construction of these ideas overlooks the rich intertextuality and complexity of life in Appalachia, the multiplicity of relationships among people and between people and their environment. In the next chapter I offer, in Niebuhr's terms, better imaginations: better ways of understanding these ideas that address that richness from an inclusive, theocentric standpoint.

"King Coal": Power and Powerlessness

In a region that has been identified as a major source of electrical power for the rest of the nation, narratives and images of power and powerlessness are abundant and complicated. As I write this, a devastating storm in the mid-Atlantic region has left communities in Appalachia without electricity for more than a week; ironically, the very people whose labor provides half the nation's electricity must temporarily survive without electricity themselves, lending a poignant concreteness to the concept of power in Appalachia. The idea of powerlessness is frequently employed to characterize the situation of Appalachians, and it is this notion that shapes nearly all other images and narratives about the region.

A STORY OF POWER IN APPALACHIA

There is a standard narrative about power in Appalachia, its nature, and who wields it over whom.[6] Before the Civil War, the people of the Southern Mountains, mainly of German and Scotch-Irish descent, were independent and isolated. In the second half of the nineteenth century, after the discovery of vast reserves of coal and the development of the infrastructure (in the form of railroads) to export it, capitalists from the rest of the nation and beyond came to the region either to develop and uplift it or to exploit it (depending on one's perspective). These outsiders took advantage of the ignorance, need, or simple guilelessness of the mountaineers and acquired the vast majority of their land through deceit, force, or legal manipulation. In addition to constructing the facilities necessary to extract, treat, and transport coal, the coal companies built housing for their workers, as well as stores, banks, and saloons, creating the notorious company town. In these towns, the company controlled every aspect of life, down to printing its own currency, "scrip," that was good only in company-owned establishments. Because of the region's economic dependence on coal, the companies were also able to establish significant, in some cases near-absolute, political influence. Thus, the early history of the coal mining regions of Appalachia is portrayed as "the establishment of the hegemony of industrial economic interests over a particularly independent, roughly equal, and relatively content enclave society."[7]

One of the most insightful and influential inquiries into the dynamics of power in the Appalachian coal regions, *Power and Powerlessness* by John Gaventa, both retells and complicates this story. Gaventa sets out to understand how the coal companies were able to maintain this near-absolute power over the people and communities—that is, to understand why the powerless offered so little resistance to their oppression. He considers two standard views of the mechanisms of power: First, power can be maintained by controlling the decision-making processes. Those with political or economic power use it directly to affect outcomes to serve their own interests. Second, power can be maintained by foreclosing the options of the powerless through force, threats, or a biased system, such that they are compelled to concede power to their oppressors. They are simply "left out" of the decision-making processes. To these conceptions, Gaventa adds a "third dimension" of power, whereby power is maintained by shaping the wants and perceptions of a community through the manipulation of symbols, myths, and language to legitimate the status quo and foster quiescence among the powerless.[8] It is this third dimension, he argues, that is involved in Gramsci's concept of hegemony.[9]

Gaventa traces the workings of this third dimension of power in the coal town of Middlesboro, Kentucky. As the town developed, the coal company brought not only industrial and commercial institutions but also a particular worldview that celebrated a modern, consumptive lifestyle and denigrated more traditional mountain ways.[10] The mountaineers had a choice, but it was really no choice at all—either a backward, archaic, inferior way of life or the modern, enlightened, consumptive world offered by industry. The dependence this created was perpetuated by churches in the company towns, which preached an "otherworldly" religion that de-emphasized issues of justice in this world.[11] As a result of this ubiquitous ideological pressure, the mountaineers internalized an image of themselves as dependent and incapable of action, such that resistance simply ceased to exist as an option.[12]

Thus, whereas a one-dimensional understanding of power (which, Gaventa argues, has dominated Appalachian studies) might attribute the lack of resistance to coal companies to some deficiency in the mountaineers themselves, and a two-dimensional view would see it as mainly a function of biased political institutions and practices, Gaventa's three-dimensional approach recognizes that consensus and quiescence are maintained by more subtle ideological forces that serve the interests of those in power. Appalachia is a colony, this conception would argue, and colonization has as much to do with controlling worldviews and imaginations as with exerting direct political, economic, and social control.[13]

Gaventa's examination of power in the coalfields adds nuance and insight to the typical story outlined above. Rather than attributing the powerlessness of miners and their communities to their own deficiencies or to the structures of power themselves, he shows that quiescence is intentionally cultivated by those in power to serve their own interests. Accordingly, rather than advocating a change in the attitudes of the mountaineers or a change in political institutions, Gaventa argues that consciousness-raising and education are required to counter the ideological colonization established by the coal industry. This is an important corrective to more traditional conceptions of power and powerlessness. Nonetheless, he leaves the main contours of the standard narrative in place; for all its insight, his understanding of power remains inadequate for this reason. Power is unidirectional, exercised by the coal industry on the "powerless" for economic benefit. The legacy of vastly unequal power relations and carefully cultivated quiescence is seen as the biggest obstacle to overcoming the current injustice in the region.[14]

POWER AND RELIGIOUS RESPONSES TO MOUNTAINTOP REMOVAL

This story, whether in its straightforward form or as nuanced by scholars like Gaventa, remains central in discussions of Appalachia and MTR, both among academics and in popular discourse.[15] Power is seen as the possession of the "coal interests" or the "ruling elite," who are generally understood to be outsiders.[16] Appalachia is described as a "sacrifice zone," a peripheral region exploited for the benefit of the rest of the nation. This is all part of the legacy of those first capitalists and their deceitful broad form deeds.[17] Sociologist Stephen Scanlan illustrates this view nicely: "This question of power, powerlessness, and exploitation cannot be ignored. A community of elites and their corporations and banks, whose primary interest is in maintaining the status quo of the economic system, uses its power to dominate think tanks and policy discussion and planning, control the media and shape public opinion, and dominate elections and governance. . . . A path of least resistance applies to mountaintop removal mining . . . because of an imbalance of power between decision makers and those whose voices dare not risk the consequences of challenging that power."[18]

Naturally, this narrative has influenced the church's thinking about Appalachia and MTR. Indeed, in discussions of MTR among Christians, as among others, it is difficult to escape the notion that the coal industry maintains absolute control over the region. Allen Johnson, a leading Christian opponent of MTR, reflects this conception when he asserts, "Coal is beyond an industry. It has a supernatural hold over the entire region."[19] An article in *Sojourners* magazine refers to the decades of domination by the coal industry, comparing it to the infamous and exploitative system of sharecropping, and the editor's introduction to the issue begins by stating, "King Coal has long ruled in West Virginia."[20] A piece in the evangelical magazine *Prism* characterizes the coal industry as "ubiquitous [and] powerful."[21]

Among Christian perspectives, this view of power finds its fullest expression in the pastorals by the Catholic Bishops of Appalachia, discussed in chapter 2. In *This Land Is Home to Me,* the bishops state, "[The saying that coal is king is] not exactly right. The kings are those who control big coal, and the profit and power which come with it. Many of these kings don't live in the region."[22] They proceed to outline the history of coal in the region in terms very similar to those in the story above. The drama is heightened in *At Home in the Web of Life,* where the story is retold in such a way that a comparison with the biblical narrative of Paradise and the Fall is inescapable.[23] Before the coming of industry, first the Native Americans and then the settlers and escaped slaves lived a simple life of spiritual harmony with nature and with one another. In the nineteenth century, however, "giant corporations" came with "outside workers," introducing social division "in rejection of God's teaching." In spite of the workers' noble efforts to unite in the face of industry's power, their strength was no match for these outside corporations.[24] In both these letters the central image is industry's near-absolute power, which remains an obstacle to justice.

To clarify this view of power, it is worth noting an alternative view that is also present in the pastorals and other Christian reflections on MTR. When the bishops turn to their vision of the future, they acknowledge that, "despite the theme of powerlessness, we know that Appalachia is already rich here in the cooperative power of its own people."[25] Here and elsewhere, the bishops recognize that power is not unidirectional, and it is not simply possessed by one class to be exercised on another class. Whatever truth there is in the standard story (and I do not deny that it expresses certain important truths about the history of Appalachia), it needs to be complemented by a view of power that is complex, nuanced, and multifaceted. Even Gaventa's version, for all its subtlety, maintains a straightforward dichotomy of oppressor and oppressed, powerful and powerless.[26] When those who are supposedly without power appropriate and internalize the worldviews and images of the powerful and thus help perpetuate them, such dichotomies become less useful. In an ideologically freighted region like Appalachia, power is not simply exercised by industry on "local communities" for the sake of economic interests. Moreover, the lives and activities of residents of the coalfields cannot be neatly characterized as either "quiescence" or "resistance."[27] As the bishops are aware, the exercise of power and agency can be more subtle and subversive than that of the coal companies or those who confront them directly.[28] I discuss the ethical and theological importance of this more complex view of power in the next chapter; my point here is simply to show that much of the discussion of MTR (including by Christians) tends to overlook it.

In addition to the Catholic bishops' statements, other denominational resolutions draw on a similar conception, either explicitly or implicitly. A resolution by the Unitarian Universalist Association states plainly, "the exploitation of Appalachia unjustly enriches other regions in the United States by providing cheap coal and thus electricity at the expense of Appalachia."[29] More than this, though, the means the churches choose implicitly express a certain view of power. By issuing national resolutions that focus on legislation and policy and call for action by national and state agencies or by the companies themselves, the national churches show a "top-down" understanding of agency and power. In Gaventa's terms, this is a two-dimensional view, since the underlying belief is that the obstacles to justice are institutional and structural.[30] If power is concentrated in the hands of the coal industry, located mainly outside of Appalachia, the only apparent recourse is to appeal to either the companies themselves or other equally external and institutional powers. Although they express a great deal of concern for communities in the region, nowhere do the resolutions by mainline denominations or the West Virginia Council of Churches appeal to the agency or knowledge of the communities themselves.[31] Allen Johnson notes that many in these communities feel "angry and snubbed" when presented with these hierarchical statements.[32]

Certainly these resolutions, and the growing national awareness they illustrate, ought to be seen as positive steps. And it seems appropriate for the national churches to speak out on MTR, since, as the Unitarian Universalist Association's statement suggests, the entire nation is complicit in this issue. Nonetheless, these statements reveal an incomplete conception of the mechanisms of power in Appalachia. A more thorough understanding would attend to the way power operates on a variety of levels and in a variety of relationships. This more complete view is present (if underemphasized) in the pastorals by the Catholic Bishops of Appalachia, as well as in the reflections of several other Christian groups and activists, although it remains a minority viewpoint.[33] I develop and defend this latter view of power and consider its specific implications for the church's approach to MTR in the next chapter.

Appalachian Identity: Insiders and Outsiders

The view of power described in the previous section, and the narrative of Appalachia's fall from paradisiacal harmony with nature to national sacrifice zone, relies heavily on the distinction between insiders and outsiders, between local communities and manipulative elites from abroad, between those who are genuinely Appalachian and those who want to oppress and exploit them. This claim is sometimes made explicitly: "Hired managers became agents of the absentee owners whose corporate offices were far removed from the coal towns themselves. This domination of the coalfields by large, often multinational, corporations . . . persists to this day."[34] Opponents of MTR argue that whatever meager benefits the practice offers go to outsiders, and the precise identification of a reviled coal executive as local or outsider, based not only on where but also on how he lives, is a matter for serious scrutiny.[35] More frequently, though, this distinction between Appalachians and outsiders is implicit in arguments about the exploitation and manipulation of the region, where phrases like "local" and "residents of Appalachia" invariably carry moral weight.

Of course, opponents of MTR and purveyors of the "national sacrifice zone" narrative are not the only ones who orient their claims around this distinction between insiders and outsiders. Those who support the practice regularly characterize their adversaries as environmentalists from elsewhere or regulators from Washington.[36] A statement released by a Massey Energy spokesman in response to protests against MTR clearly illustrates the power of the insider-outsider dichotomy:

> It is my understanding that all but one of the fourteen protesters who were arrested for scaling a dragline at Massey Energy's Twilight Mine in Southern West Virginia are residents of states other than West Virginia, such as Maine, Oklahoma, Michigan, and Florida. As a native of Appalachia and a resident of West Virginia, I find it hypocritical that these folks come from out of state to waste West Virginia's tax dollars [by necessitating a police response]. . . . It is clear that these folks are not concerned with the people, the environment, or the economy of West Virginia. Their efforts are purely about gaining hype and media attention for their out-of-state funders. . . . It is time for all West Virginians to stand up and say enough is enough to the protesters. We know our state and our economy and we won't be told what to do.[37]

THE "REAL" APPALACHIA

The clear implication in all applications of this dichotomy is that the people of Appalachia know what is best for the region and have the right to decide for themselves. An underlying assumption, of course, is that all *real* Appalachians must feel the same way about MTR (whether for or against it); therefore, anyone who disagrees must be an outsider (at least in spirit) and must have ulterior and undoubtedly insidious motives (such as "gaining hype and media attention for their out-of-state funders").

There is, however, a more profound discourse operating within this dichotomy, a discourse about who and what can be considered authentically Appalachian and who may speak for Appalachia. Claims of legitimacy as insiders are not primarily about geographic boundaries—where one (or one's kin) was born or raised; they are about identity. When they mobilize the dualism of insiders versus outsiders, participants in contemporary debates trade on the notion—long established in discourse about Appalachia—that there is a coherent class of "Appalachians" and that the members of this class share certain characteristics and common interests; at the same time, they challenge past portrayals of this class as inaccurate in some way. Sociologist Rebecca Scott contends that the debate about MTR, and about the future of coal mining more generally, is really a struggle between different conceptions of this Appalachian identity as it relates to an imagined American identity that has continually denied its validity.[38]

The people of Appalachia have long been an enigma to the larger culture: like the mountains themselves, they have been understood as both part of the nation and separate from it. This paradoxical otherness is, of course, a political creation, constructed to serve the interests of one group over another.[39] As Batteau argues, in the case of Appalachia, this construction of otherness took place as the nation was seeking to solidify its identity based on the image of the frontier and the conquest of nature, and as the conflict between North and South was becoming increasingly tense.[40] In this context, the image of a kind of internal frontier, where nature was still continually being conquered by heroic means, could be mobilized for political purposes. Considered both outside of and more pure (because closer to nature) than the dominant American culture, Appalachia was used to criticize the Northern establishment; simultaneously, because it was geographically in the South yet innocent of that region's particular vices, it was also a critique of Southern slaveholders. Thus, in the period leading up to the Civil War, the basic image of Appalachia was already in place: a region whose relationship to nature represented the ideals of the nation, yet whose peculiar culture and traditions set it apart as backward.[41]

From that time forward, this basic pattern was modified and reconstructed for diverse ends. Historian Henry Shapiro describes how Protestant denominations constructed the region as the ideal mission field by emphasizing its paradoxical otherness and Americanness. The southern mountaineers were different enough to require missionary outreach but similar enough to place a special claim on US churches.[42] This paradox of similarity and otherness continued to shape the nation's attitude toward the Appalachian region.[43] At the same time, in light of a growing sense of a unified national identity, this otherness needed to be explained. Ethnicity (focusing on the dominant Scotch-Irish heritage), environment (the isolation created by the mountains), and lack of community were offered as explanations.

According to Scott, this politically constructed notion of Appalachian difference makes the economic exploitation of Appalachia possible.[44] Such exploitation is conceivable only because Appalachia and its people are understood in the national consciousness as different, simultaneously ideal and deviant; they are viewed with romantic nostalgia and bemused derision, but always as decidedly other. "These epistemologies of disgust and social distance," Scott argues, "help create the conditions of the possibility for some of the most dangerous environmental exploitation in the United States and the designation of Appalachia as a sacrifice zone."[45] At the same time, the construction of mining as normatively masculine (in a traditional sense, emphasizing provider status, toughness, and progress) and white creates a context in which MTR makes ideological sense (even though, she argues, it does not make economic or environmental sense).[46] Here, we can see the continuation of the theme of the heroic conquest of nature that was part of the original "invention" of Appalachia. While the region is set aside as different, somehow deviant, and sufficiently distant to avoid analyzing its realities too carefully, mining is constructed as the quintessentially American (which is to say, white and masculine) heroism that subjugates nature to human needs, and MTR is this heroic conquest at its most extreme. From the perspective of its supporters, MTR is the means by which Appalachians can prove their true Americanness to the nation that has so long denied it. Ironically, though, the poverty, physicality, and abjection associated with mining distance miners and Appalachia from the stereotypically American values of property, development, and independence.[47]

IDENTITY AND PLACE

The debate over mining and its future in Appalachia is, fundamentally, an attempt to negotiate the relationship between Appalachian identity and the larger American consciousness. According to Scott, at the heart of this struggle are two competing understandings of place. The mining industry views the land as "empty, useless, and waiting to be improved." Industry supporters hold an abstract view of space that is shaped by capitalist ideology.[48] This is especially true of MTR, she says, where a place is literally destroyed for the sake of the resources beneath it. Meanwhile, those who oppose MTR—specifically, those who adopt an environmental justice approach— offer an alternative view that is concretely situated in particular places and a particular cultural context.[49] Whereas MTR supporters can argue that a grassy plain where a mountain once stood is an improvement, its opponents defend the irreducible value of that particular mountain.

This, too, is a familiar notion. A connection to place has been used to characterize the peculiar nature of the people of Appalachia more than any other factor; the mountains and their inhabitants have always been identified with each other.[50] The otherness of the Appalachian people is reflected in the otherness of the Appalachian Mountains: rugged and sinister, romantic and bucolic, untamed and free. As noted earlier, in a nation whose identity was being forged around both the romance and the conquest of nature, this close association with a wild place was ambivalent, and it continues to be so.

Scanlan provides a good example of the continuing power of this notion of Appalachians' connection to place: "Perhaps most damaging to some is the loss of emotional and spiritual attachment to the mountains that the Appalachian citizens call home. With strong historical, family, and cultural ties to the land that have been reinforced generation after generation, there are real emotional scars left on individuals."[51] Or, as Scott puts it, "MTR is fundamentally, basically, opposed to the coalfield cultural tradition of attachment to place."[52] As these quotations illustrate, the idea of Appalachians' close association with their land can be, and has been, mobilized to oppose the exploitation of both people and land. At the same time, this notion is part of the logic that makes exploitation possible. As Scott points out, when land and people are conceptualized as problematic, as resistant to modern development, they become expendable in the interest of progress.[53] The problem, therefore, is that even when the idea of Appalachians' connection to place is used to oppose the arguably exploitative practices (such as MTR) that seem to sever that connection, this idea perpetuates the same imagination of the region that sets it and its people apart as curiosities. If, as Scott argues, exploitation is possible because representations of Appalachia, "whether . . . in valorized or debased form," erase its internal heterogeneity and establish it as fundamentally different, then any appeal to the seemingly self-evident Appalachian connection to place needs to be carefully examined.[54] In the final chapter, I return to this problem of place and defend a conception of the connection to place based on a carefully cultivated relationship, rather than on innate identity.

THE POLITICS OF IDENTITY IN APPALACHIA

Batteau describes what he calls "Appalachia-building," or the movement to "uplift" Appalachia, in the 1970s.[55] At the time, this movement was propelled by the environmental devastation of strip mining, the precursor to MTR, and Batteau's description continues to resonate in the context of the current debate. He points to three ideological strands that depicted Appalachia variously as the home of "contemporary ancestors, country cousins in need of uplift; . . . [as] a wild land . . . populated by inconsequential savages; . . . [or] as the locus of perfect victimization."[56] Each of these constructions was an effort to discern an authentic regional identity as the first step toward political mobilization. Each represented a cooperative effort by interests both inside and outside the region, yet each sought to establish a boundary around "what is truly Appalachian," portraying any deviations as exogenous in origin, and then to represent that true identity to a national audience.[57]

Batteau's three strands may be overly simple. Certainly, when we consider contemporary debates over Appalachian identity, it is clear that while these images retain their salience, they have been modified and blended together. Appalachia remains a "locus of victimization," described now as a "sacrifice zone." The image of "contemporary ancestors" can be found in Scanlan's invocation of "strong historical, family, and cultural ties to the land that have been reinforced generation after generation." The image of "savages" has, of course, been jettisoned, but its modern-day equivalent is seen in comparisons of Appalachia to the "third world," a locution that Scott points out reaffirms a modern, racialized conception of development and places Appalachia securely outside of that conception.[58] The contemporary categories stand in continuity with the earlier ones. In the debate over MTR, claims of insider status and authentic identity can be understood as concerted attempts by people both within Appalachia and beyond it to control how the region's identity is represented to the larger society. Moreover, what Batteau affirms about the earlier ideological strands is still true: these attempts are the latest in a long tradition of politically motivated constructions of Appalachian identity. Any effort to assert an Appalachian identity "must come to terms with the image and reality of Appalachia created by its patrons and forebears; until it does, it will be suspended in webs of significance it has not spun."[59] The assertion of a regional identity may be, as Batteau argues, the first step in political mobilization; yet it is not, he says, without its ambiguities and traps.[60]

APPALACHIAN IDENTITY AND RELIGIOUS RESPONSES TO MOUNTAINTOP REMOVAL

As with narratives of power and powerlessness, the church's response to MTR has incorporated this dichotomy between insiders and outsiders and the attendant images of Appalachian identity. The Unitarian Universalist Association's statement that "the exploitation of Appalachia unjustly enriches other regions in the United States by providing cheap coal and thus electricity at the expense of Appalachia," for example, trades on the moral weight of the division between locals and outsiders.[61] A statement by the United Methodist Church more subtly invokes the idea that Appalachians have a special connection to their land and to the past: "mountaintop removal mining, by destroying home places, is also destroying ancestral ground, sacred ground where generations after generations have lived, gone to church, married, made and birthed babies, taken family meals, slept in peace, died and been buried."[62] The Presbyterian Church (USA) makes reference to several established images of Appalachian identity: the people's special relationship to the land and a nostalgic association with the bucolic past (as evidenced by use of phrases such as "roots," "family home place," and a "lifetime of memories" from which people have been exiled), and Appalachia as a sacrifice zone (mining "destroys the beauty and productive capacity of the land . . . eliminating future or alternative economic opportunities" and creates "a cycle of poverty").[63]

In addition to invoking some of the tropes used to define Appalachia throughout its history, these denominational resolutions are noteworthy in other ways. Just as their calls for action by state and national governments or by companies express a certain view of power, the genre of national denominational pronouncements on the injustice of MTR seems to reveal either ignorance of or lack of regard for the long history of the construction and manipulation of Appalachian identity by and for the nation as a whole. I am not suggesting that the churches willfully distort the reality of Appalachia; as noted earlier, however, much (perhaps most) of the political construction of Appalachian identity has ostensibly been well intentioned. These denominational statements continue, in their own way, the well-established tradition of representing Appalachia to and for a national audience for particular political ends. In doing so, they also play into many Appalachians' expectations of outsider interference, as reflected in Massey Energy's dismissal of "outside environmentalists." One person working to educate church leaders about life in Appalachia acknowledges that resolutions like these can forestall real dialogue, since people living in the area often feel that their voices are being ignored and therefore choose to withdraw from the debate.[64]

The narrative of the Fall recounted in the Roman Catholic pastorals exemplifies these assumptions about Appalachian identity. Familiar images of Appalachia and Appalachians—a spiritual harmony with nature, a nostalgic premodern folk culture, a combination of fierce independence and close community and kinship ties—are prominent throughout the letters.[65] The bishops place commendable emphasis on listening to "the voice of the region"; yet the pastorals minimize the heterogeneity of those voices, relying instead on established conventions. Threats to "Appalachia's old traditions" are seen as coming from an alien culture—the "modern consumer society."[66]

Again, the purpose of assessing these resolutions' reliance on long-standing constructions of Appalachian identity is not simply to criticize them or deny their importance. Rather, it is to engage in the Niebuhrian process of self-examination described in chapter 3, since, as Batteau confirms, these images have always been ambivalent and problematic (even as they may also be quite useful for political or moral action).[67] Unsurprisingly, many individual religious activists who are directly involved in this debate show a greater awareness of this ambivalence than is reflected in the statements of the larger church bodies. Some of the conventional images are invoked, especially by those who were involved in or influenced by the Roman Catholic pastorals. Carol Warren, a Roman Catholic activist with the Ohio Valley Environmental Coalition, cites the severing of a close connection to place and the loss of home places as one of the most serious effects of MTR.[68] Father John Rausch, director of the Catholic Committee of Appalachia, refers to the region as a "mineral colony," indexing the image of Appalachia as a "sacrifice zone."[69] At the same time, these activists and others are well aware of the heterogeneity of the people living in Appalachia and of the need to avoid easy dichotomies and polarizing characterizations that foreclose real understanding. Parish ministers, in particular, are cognizant of the variety of cultures and interests that make up the region and of the need to balance prophetic witness with respect for this diversity.[70]

One prominent activist, author and Episcopal deacon Denise Giardina, reflects this strategy of both drawing on and modifying conventional images of Appalachia. In a speech given to the ecumenical organization Christians for the Mountains, Giardina distinguishes between those who work the mines, who are either "desperate for jobs" or fooled by the "American dream of conspicuous consumption," and the "people who control the companies," whom she identifies as the real enemy.[71] Thus, within the typical binary of exploitative outsiders and Appalachian victims, Giardina (like Gaventa) acknowledges the additional factor of local participants' complicity in the exploitation. At the same time, though, she argues that those who control the industry are "bent on destruction . . . cut off from community . . . cut off from beauty . . . [and] cut off from God." By emphatically separating those who control the industry from notions of community and beauty, which have long been connected with Appalachian identity, she portrays the controllers (presumably, owners and politicians) as being truly outside of Appalachian culture. Note, too, that the participation of those who work the mines is explained (if not excused) by either their desperate poverty or their ensnarement in the larger (external) culture. Both of these index familiar tropes of Appalachia.

Claims to insider status are involved in a long tradition of representing Appalachia to the nation. Like any discourse, this tradition is neither purely good nor purely bad; nor are the images it employs necessarily true or false. Rather, this representation simplifies a complex, heterogeneous reality, and as such, it has had problematic effects. When activists on either side of the MTR debate invoke the dichotomy of insiders versus outsiders or the stereotypical images of Appalachian identity that have informed that dichotomy, they participate in what Batteau calls "webs of significance [they have] not spun." Although some notion of Appalachian identity may be inevitable and may even have great efficacy, responsible engagement of this issue requires that such notions be employed carefully and strategically. How this might be done, along the theocentric lines already developed, is discussed in the next chapter.

Changing Places: Destruction and Reclamation

The conceptions of power and powerlessness and of insiders and outsiders, as they are mobilized in the debate around MTR, have complicated and ambiguous histories. Although both these conceptions point to important realities and may therefore be inevitable and valuable, to invoke them uncritically obscures the dynamic, multifaceted social and political processes behind them, as well as the rich heterogeneity of the mountains. Here, I advance a similar argument about the concepts of destruction and reclamation. Rather than being straightforward descriptions of processes involved in MTR, these terms involve a variety of interrelated and politically driven claims. Their uncritical application (or, in the case of reclamation, uncritical omission) is question-begging, in that it takes for granted precisely what needs to be examined from a moral standpoint: how does MTR change a place, and how can or should these changes be mitigated or avoided?

DESTRUCTION AS A POLITICAL CLAIM

At first glance, this third pair of terms seems far less problematic than the first two. Unlike notions of power or notions of identity, the notion of destruction seems to involve a relatively concrete and straightforward reality. It is likely that anyone looking at an active MTR site would agree with this assessment: the destruction of the mountaintop seems complete and undeniable. And indeed, for most opponents of MTR, the fact that it destroys mountains is self-evident. One important work on MTR opens with large-format photographs of active mine sites under the words "destruction," "devastation," and "desecration" in enormous type.[72] The same book features a chapter entitled "Mountaintop Removal: The Destruction of Appalachia."[73] Wendell Berry describes the practice as "the total destruction of the land and the land's communities."[74] Even Rebecca Scott's careful analysis of the meaning-making dynamics at work around MTR asserts that "MTR is simultaneously destroying forests and ecosystems [and] flattening the beautiful Appalachian Mountains" and is "annihilating the place."[75] These are just a few examples; the claim that MTR is incontrovertibly and absolutely destructive is axiomatic among anti-MTR activists.

Among the religious opposition to MTR, this claim becomes virtually an article of faith. A common theme is that MTR destroys God's creation and is therefore blasphemous. One prominent activist, Larry Gibson, cites Revelation 11:18—"The nations raged, but your wrath has come, and the time for judging the dead and for destroying those who destroy the earth"—to illustrate God's wrath against those responsible for MTR.[76] Kathy Selvage, former vice president of the Virginia nonprofit Southern Appalachian Mountain Stewards, states plainly, "If you believe there is a God, and you believe that he created this earth, then every time you blow up a mountain, it's like slapping his face."[77] Denise Giardina proclaims that mine owners and operators are "bent on destruction. Destruction is their life's work." Consequently, "they are cut off from God."[78] The denominational resolutions generally opt for more specific descriptions of the demolition or removal of mountains and forests, although the language of "destroying the environment" is present in some of these statements.[79] For Christians involved in the fight against MTR, the complete destruction it causes is not only undeniable; it is blasphemous.

Far from being self-evident, however, destruction is an ambiguous and politically constructed notion that comprises a variety of questions and claims, as political ecologists remind us. Even a straightforward definition of the term—for example, "the substantial decrease in either or both of an area's biological productivity or usefulness due to human interference"[80]—leaves a variety of questions unanswered: Decrease compared to what? What sort of productivity or usefulness? How is human interference assessed? Does the decrease have to be permanent? Responding to these questions involves choosing among criteria based on particular interests.[81]

Even when parameters exist for determining some of these conditions (for example, measuring a decrease by comparing an area's productivity with that of a similar area or by comparing the same area's productivity in past years), certain judgments remain necessary. Frequently, a decrease in one sort of productivity is accompanied by an increase in another. In addition, what counts as an appropriate sort of "usefulness" in a particular region is a profoundly political decision.[82] In the case of MTR, this can be seen in supporters' claims that the level grasslands produced by reclamation are, in fact, more useful than the mountaintop was, or that these grasslands provide habitats for different and even rarer wildlife. Proponents of MTR also point out that similarly dramatic processes are used to build roads and other human edifices in the mountains, yet these are not as vehemently decried as "destruction," presumably because their usefulness is more generally accepted. Besides the question of usefulness, the effects of human interference are not so easily determined, since natural systems are highly dynamic. Measuring the impact of human action against such a variable standard can be a challenge.[83]

Certainly, in any particular case, some of these questions are more compelling than others. With MTR, usefulness and productivity remain open questions for some, but the question of human interference brooks little debate. Whatever natural dynamism exists, the impact of human actions on the environment is readily measurable, whether in tons of overburden displaced or miles of streams buried. Yet even after acknowledging human interference and answering the question of productivity (for example, advancing what many would see as the uncontroversial argument that a forested mountaintop and free-flowing streams are more productive in a variety of ways than the "moonscape" of an active mine site), the applicability of the term "destruction" is still not incontrovertible. A degree of permanence is implied in the meaning of destruction.[84] To argue that an ecosystem has been destroyed suggests that recovery is impossible, or at least that it would require such a long span of time as to be virtually impossible. Thus, in the case of MTR, examination of the notion of destruction necessarily involves a consideration of reclamation.

THE QUESTION OF RECLAMATION

The claim that MTR is destructive entails the belief that what is lost—the productivity or viability of forests, streams, or soil—can never be regained. And if opponents of MTR see its destructiveness as self-evident, the belief that reclamation is impossible is equally axiomatic. The statement by the Presbyterian Church (USA) is the only one that refers to reclamation at all, and it is illustrative: "Streams, mountains, and forests damaged by mountaintop removal coal mining *can never be restored* to support the community of life that God created."[85] As noted in chapter 1, a variety of claims have been made about the possibility of real reclamation—the restoration of a functioning mountaintop ecosystem. Publications from the West Virginia Coal Association and the University of Kentucky show lush sloping forest landscapes that were once surface mines.[86] Yet opponents of MTR see this as deceptive propaganda and emphasize that most so-called reclamation involves replacing hardwood forests and mountain peaks with desolate grassland.[87] Meanwhile, the forestry reclamation approach promises productive forests at less cost to mine owners, but research shows that while reclaimed forests and streams may reproduce some functions of the original ecosystem, they do so at a lower level (see chapter 1).

Some Christians elevate the arrogance and deception of reclamation (like destruction) to blasphemy. When supporters argue that MTR improves on nature by creating more useful land (often exemplified by Twisted Gun Golf Course, regional airports, or the Earl Ray Tomblin Industrial Park[88]), many Christians interpret this as an affront to the mountains' creator. At least one mountaintop miner agrees with this interpretation, but he denies that coal mine operators are making this claim: "It would be blasphemy if they said they were going to make this land better than what God made it. All they are saying is that they are making it more useful."[89] Perhaps nothing better expresses Christian resistance to the notion of reclamation than the sign posted by one activist: "God was wrong: support mountaintop removal."[90]

Of course, the determination of what constitutes reclamation is as selective and as politically constructed as the notion of destruction. Even with minimal or no human intervention, ecosystems recover from human impacts in a variety of ways.[91] Some systems may be relatively resilient, whereas others may be more fragile; still others may show resilience to a low level of human activity but are unable to recover from higher levels of human impact. In some ecosystems, recovery may naturally lead to a new state rather than simply reverting to the original state; thus, even if an impact is not "reversible," recovery in some form may still be possible.[92] Again, these variations in the recovery of natural systems occur without any human intervention toward reclamation.

These observations about environmental recovery in general are relevant to MTR. Surface mine sites that predate the Surface Mining Control and Reclamation Act of 1977 are home to mature hardwood trees and show significant signs of recovery.[93] In other words, despite minimal human-led reclamation, these sites have recovered well, suggesting that mountain ecosystems exhibit a degree of resilience. But these sites are now forty years old or more. What is a permissible time frame for recovery?

Like the definition of destruction, the goals of reclamation are not easily established. Environmental assessments of mining and reclamation plans tend to view the land as wilderness prior to mining, and the goal of reclamation is seen as a return to wilderness in some form.[94] People on both sides of the MTR debate echo this assumption in different ways: proponents argue that the land was "just sitting there," useless, before mining and reclamation, and some opponents want the land to be preserved as pristine wilderness.[95] Yet the mountain people have been using this land for recreation and sustenance since well before the European settlers arrived, and current inhabitants continue to do so, treating the land as a local commons.[96] Returning the land to some constructed image of wild nature, without taking into account the usage patterns of local residents, is an illusory goal, whether advocated by proponents or critics of mining and reclamation.

Defining reclamation involves these and other judgments and decisions. On a technical level, it requires a careful assessment of the various factors necessary for a particular environment to function, from plant and animal species to water conductivity. But like the idea of destruction, reclamation involves judgments about what kinds of use or productivity are preferred and what makes a particular ecosystem valuable. Regardless of the apparent absurdity of a golf course on a leveled mountaintop or a vineyard on a reclaimed mine site,[97] decisions about what constitutes appropriate use are not foregone conclusions. The issue of reclamation also necessitates an awareness of the dynamism of natural systems. The goal of reversion to the "original" or "premining" structure and function of ecosystems has always been a moving target.[98] Most important, perhaps, is the recognition that humans unavoidably interact with and change their environments and that careful consideration of which changes are acceptable and which are not is more helpful than the insistence that any particular change is irrevocable.

CHANGING PLACES

Permanence is not a characteristic of the natural environment, even in a region as old as the Appalachian Mountains. In addition to the region's natural dynamism, humans have altered the mountains in a variety of ways for more than a century. Most of the forests that are being razed by MTR are second-growth forests that replaced what was clear-cut by the timber industry in the nineteenth century.[99] Before, during, and after mining, residents use the land in ways that may have more or less permanent impacts. Change is a fundamental characteristic of human interaction with the environment. As Paul Robbins states: "Systems are driven to new states, some recover, others don't; some take new forms, which in turn enter new states, or return to earlier mixtures of elements." Given these processes, he suggests that it is more useful to think about human impact in terms of transitions rather than "broken or fixed" environments.[100]

Ultimately, destruction and reclamation are two ways of describing the changes people impose on the land. Each of these terms involves a number of interrelated decisions about usefulness and productivity, temporal and physical scale, and the appropriate relationship between humans and their environment. These terms—like the others discussed in this chapter—undoubtedly have significant value for naming and transforming real actions and attitudes. Nonetheless, they are relatively blunt instruments, and their uncritical application may obscure more than it clarifies. If the notion of destruction and its correlate notion of reclamation imply a departure from or a return to a pristine wilderness that is untouched by human intervention, then these terms ignore the way, as Robbins puts it, humans "produce" nature both ideologically and materially.[101] In particular, like the concepts of power and powerlessness and insiders and outsiders, the concepts of destruction and reclamation tend to overlook the complex relationships within and among the Appalachian people and between these people and their land. Therefore, the mobilization of these ideas in the debate over MTR needs to be reconsidered and modified to reflect this complexity. Rather than simply labeling MTR destructive and reclamation a fantasy, we should consider the kinds of human-driven changes occurring in Appalachia, discern which changes are acceptable and which are not, and find ways to prevent or miti-gate what are deemed to be unacceptable changes.[102]

Understanding the Meanings of the Mountains

As Scott argues, the debate over MTR is ultimately a struggle between differ-ent understandings of place. Her judgment that proponents of the practice enact a view of place that is abstract, empty, and fundamentally opposed to the typical Appalachian attachment to place, however, is guilty of the same oversimplification that characterizes other discourses about MTR. Rather, a wide array of perspectives are operative, and there are a variety of ways to understand the meaning of the Appalachian Mountains. As Niebuhr's discussion of value makes clear, it is not true that some people value the mountains themselves, while others value them only for the resources they provide. The mountains may have a variety of meanings and a variety of values for any given person. A mine owner who takes over abandoned and unreclaimed surface mines, reclaiming them to his own high standard, enacts his own understanding of place that does not fall easily into Scott's dichotomy. People in Appalachia, like people anywhere, "construct a sense of place out of the flotsam and jetsam of many influences—whether soap operas, sermons, country music, layoffs, advice from the *Farmer's Almanac*, etc., etc. . . . [and] examination of these constructions is essential if we are to locate Appalachian people in their rightful roles."[103]

The concepts of power, identity, and destruction explored in this chapter are part of this process of construction. These narratives work together to shape how people understand Appalachia, not in simple either-or binaries but in a variety of interrelated ways. The dualisms of power-powerlessness, insider-outsider, and destruction-reclamation simplify the complex real-ity of these relationships, just as Scott's dichotomy of place-based versus abstract space does. The questions that motivate Scott's study illustrate the connections between these categories (especially destruction and identity) and relations to place, as well as the limitations of these simple dichoto-mies. She asks, "How can people simultaneously love a place and support its destruction? How can two groups of people see the same landscape in entirely different ways, as a wasteland or an Eden, as an abomination or a site of technological improvement? How do people reconcile MTR with the famous Appalachian attachment to place?"[104] These attitudes seem less clearly contradictory when they are viewed in light of the complex social construction of these categories. Notions like "destruction," "wasteland," "abomination," and "Appalachian attachment to place" conceal much more complicated and dynamic attitudes.

Nonetheless, as I have acknowledged throughout this chapter, these ideas remain useful and necessary for articulating certain realities, and as such, they have strategic value. In keeping with the Niebuhrian theocentric approach, the goal is not to discard these terms and the discourses that go with them but rather to explore them carefully—to understand precisely "what is going on" in these discourses and employ them more responsibly. The goal is to challenge what Niebuhr calls evil imaginations and, specifically, the church's participation in them and to replace them with concepts and discourses that reflect more clearly the inclusive values and purposes of God the divine valuer.

Thus, in this chapter I have examined the complicated and, in some ways, inadequate imaginations that have shaped the debate about MTR. In particular, I have argued that the notions of power and powerlessness, insiders and outsiders, and destruction and reclamation are part of a long history of the political construction of Appalachia—a history that has frequently overlooked the complex relationships and dynamics of life in the region. If the church is to avoid this pitfall in its discussions of MTR, it must challenge and modify its use of these narratives with images that are more inclusive and theocentric. Suggesting what forms these narratives might take is the focus of the next chapter.

5

All My Holy Mountain

Power, Identity, and Reclamation from a Theocentric Perspective

The effort to respond ethically to MTR has caused us to reconsider some of the problematic and complicated discourses that have given shape to the church's (and society's) understanding of this issue. In other words, it has led us to examine the imaginations that Christians and others use to interpret MTR and shape a response to it. What has been excluded from these imaginations is the complexity and heterogeneity of the lives of Appalachians.

A close examination of key discourses in the debate over MTR is revelatory. Michel Foucault has shown that the liberation and examination of subjugated knowledges, of the historical knowledge of struggle, can illuminate the effects of the power exercised by dominant discourses. That is, attention to these marginalized discourses challenges the unity and exclusivity claimed by dominant discourses and shows them to be attempts to shore up political and economic power.[1] Put differently, as discussed in chapter 3, counterhegemony and countermemory, in their complexity and intertextuality, challenge hegemonic imaginations by showing reality to be multifaceted and truth claims to be exercises of power. Thus, insofar as our examination of the typical narratives of power, identity, and destruction in Appalachia exposes their political nature and reveals the multivocal realities they exclude, it relativizes those narratives and provides a more complete understanding of the issues they address.

Thus far, this process of critical examination is in accord with Niebuhr's concern for the irreducibly relational nature of moral decision making and his insistent attention to discerning "what is going on." A responsible, theocentric approach to ethics demands careful interrogation of our interpretations and their deficiencies, since these interpretations shape our responses.

Yet a theocentric approach goes even further: Inadequate imaginations are destructive not only because they are incomplete. More important, they are destructive—even evil—because they locate finite human selves and interests at their centers.[2] Thus, these imaginations are scrutinized not only to see their lacunae and deficiencies but also to rehabilitate and reconstruct them theocentrically. For all the complexity and relativity of human values and worldviews, imaginations can finally be better or worse, more or less God-centered. The goal of my theocentric approach, then, is not only to relativize human interpretations and imaginations but also to articulate more God-centered ones based on what can be said, with due humility, about God's values and purposes.

In this chapter I describe theocentric perspectives that reshape imaginations of power and powerlessness, insiders and outsiders, and destruction and reclamation. For each of these ideas, I begin by summarizing what I take to be a theocentric view of the issue, explaining it with reference to Niebuhr's work. These italicized statements represent my own extrapolation of Niebuhr's position. Consideration of the theocentric view supports the argument, begun in the previous chapter, that in some cases, the mobilization of these ideas represents inadequate imaginations. I defend the applicability of the theocentric view to MTR and Appalachia by indicating how this perspective resonates with a variety of other perspectives on Appalachia. Finally, because the interpretive framework not only shapes but also is shaped by responsive actions—that is, reflection and action are mutually informing—I suggest some practical implications for Christian responses to MTR, offering specific examples from this region and this issue.

The overall shape of these reflections describes an approach to MTR characterized by the typically Niebuhrian values of humility and confidence: humility because human actions, for better or worse, are finite and ambiguous, and confidence in the benevolence of the divine power behind creation (however inscrutable it is). Within the limits of these general attitudes, we can also attempt to discern what the purposes and interests of God might be, or at least how these purposes and interests might change our approach to MTR. As in James Gustafson's *Ethics from a Theocentric Perspective* (described in chapter 3), this may involve more inclusive relevant wholes, that is, a more expansive view of the persons, communities, and values worthy of moral consideration. It may also incorporate the insights of other disciplines, such as the social sciences, to the extent they clarify God's action in the world. Most fundamentally, however, this view insists that whatever the problems and limitations of humanity's best (and worst) efforts to care for Appalachia, they remain God's holy mountains.

Power and Hope

Power ultimately lies with God and is exercised in and throughout God's creation, often in unexpected ways. Whether he is discussing an abstract object of confidence or the revealed God of Christian history, Niebuhr's thoroughgoing monotheism naturally tends toward a view of power as a property of God. Throughout his work, Niebuhr describes an ethic whereby an agent displaces himself as the center of his worldview and assumes an attitude of dependence on the ultimate power; he sees this power at work in all things and trusts that this power is benevolent. At the same time, the Christian story reveals that this divine power is expressed more in weakness and suffering than in might and triumph.

At the foundation of Niebuhr's ethic is the response to "the radical action by which I am," which may be either a response of distrust or one of trust and confidence in the benevolence of this creative power.[3] The fundamental decision to trust in the goodness of the ultimate power behind all being is, in Niebuhr's understanding, faith in God, since the idea of God is essentially the idea of goodness and power together.[4] This response conditions our interpretations of all other actions: "Responsibility affirms: 'God is acting in all actions upon you. So respond to all actions upon you as to respond to his action.'"[5] One way of articulating the basic attitude of theocentrism, then, is that the creative power behind all being is good, and this power is the agency behind the actions of all finite beings. This attitude is seen paradigmatically in Christ, who understands, even in the face of death, that "Pilate would have no power . . . had it not been granted to [him] from above."[6] According to this view, all powers that operate in creation are ultimately exercises of God's power—a power that human beings can trust, even as they cannot comprehend it.[7]

The responsible self in search of coherence trusts that the absolute power, which is expressed in all finite powers and actions, is fundamentally good. At the same time, however, Christian revelation challenges all intuitions about the precise nature of that power. In Christian history, the ultimate power manifests itself primarily in the person of Jesus Christ and in his death and resurrection.[8] The life-giving and death-dealing power is revealed in the weakness of one who suffers and dies, and it is revealed as being more powerful than death. It is a power that is "made perfect in weakness and . . . exercises sovereignty more through crosses than through thrones."[9]

The theocentric ethic calls for the reevaluation of inadequate or destructive imaginations in light of God's revelation, replacing humans and human interests and ideas at their center with the interests and purposes of God. In this case, the view of power articulated here suggests two emphases for this reevaluation. First, it decenters humans and human power from our understanding of power, arguing that whatever power is exercised in creation begins ultimately with God. Second, it displaces human ideas about power, arguing instead for a power that looks like weakness.

A HIGHER POWER AND A LOWER POWER

Applied to Appalachia, this view of power challenges the conventional view described in chapter 4 by calling attention to the power that is both above and, in a sense, below the power of the coal industry. Recall that the conventional view, in its various forms, tends to portray power as resting completely in the hands of the coal industry, which exercises near-absolute control over all aspects of life in the coalfields. The current struggle against MTR confronts the same monolithic power that created the company towns. Considered in light of the theocentric view of power, it is clear that this represents an inadequate imagination, in that it places human ideas and interests at its center, viewing power as a human possession and describing it in the worldly terms of political and economic power. The theocentric view, in contrast, emphasizes the power of God that is exercised throughout creation. This means that all exercises of power are expressions of that unified power behind creation; conversely, it means that divine power is enacted in an infinite number of finite relationships and actions. Power in the world is not monolithic; it is not possessed by one group to be imposed on another. It is multidirectional, operating in the innumerable actions of creation. Moreover, as the radical understanding of power expressed in Christian revelation makes clear, if one wishes to see the exercise of divine power, one must turn to those who are apparently weak—the "powerless."

This counterintuitive challenge to the conventional view finds support in other arenas. Influenced by postcolonial discourse, many scholars of Appalachia complicate the traditional narrative, finding power in unexpected places. Stephen Fisher, in an examination of the history of so-called resistance efforts in Appalachia, concludes that these efforts are more nuanced and subtle than simple narratives of oppression and resistance would suppose, constituting a "web of both resistance and complicity."[10] In this context, he argues, what is needed is a search for transformative elements throughout the culture, in unexpected places. Rebecca Scott contends, with Foucault, that rather than being simply repressive, power is generative and creative, such that an "oppressed mountain woman," constructed as dependent and marginalized, "has the potential to become a radical activist, mobilizing her love of a place and her nurturing of a family in an intellectual, moral, and political struggle."[11] Geographer Stephen Hanna contends that power and resistance, or hegemony and counterhegemony, are mutually implicated and continually renegotiated: "the existence of resistance within hegemony and hegemony within resistance, ensures the continued reproduction of alternative representations and meanings."[12] While these scholars (and others) draw on a variety of theoretical approaches to explain the dynamics of power, and while they obviously do not share Niebuhr's theocentric commitments, what they have in common is the realization that power is exercised and challenged in a multiplicity of relationships, rather than unidirectionally from the "powerful" to the "powerless."[13] Complicity and participation in oppressive exercises of power reach far beyond the usual hegemonic suspects, and transformative resources of power may be found in unlikely places. Narratives of power and powerlessness, or of colonization and sacrifice zones, may overlook these dynamics.

Many of those who challenge MTR from a Christian perspective are aware of this complexity. As noted in chapter 4, even though they describe oppression and power in somewhat monolithic terms, the Catholic Bishops of Appalachia acknowledge that "despite the theme of powerlessness, we know that Appalachia is already rich here in the cooperative power of its own people."[14] The bishops point to the creativity in local communities and churches and confirm the possibility of transformation, "for it is the weak things of the world which seem like folly that the Spirit takes up and makes its own."[15] Other activists share a similar confidence that God's power is at work in unlikely places. Lon Oliver, executive director of the Appalachian Ministries Educational Resource Center (AMERC), argues that the church's role in a debate like MTR is to listen to those involved and find out "what God is up to" in a given situation. This listening, he believes, can lead to "life-giving" discussions that call attention to both shared complicity and truly transformative action.[16] These attitudes, shared by others in the region, illustrate an awareness of power that is both above and below that of the coal industry: a divine creative power that finds expression in unexpected and apparently weak or "powerless" places.

IMPLICATIONS OF THE THEOCENTRIC VIEW: HOPE AND HUMILITY

This theocentric view of power leads to an approach to MTR characterized by both humility and hope, because it affirms that power ultimately belongs to God, who "ministers indeed to all our good but all our good is other than we thought."[17] This affirmation of God's power instills humility, since it displaces human interests and agency from the center of our moral decisions. Human power is finite power, and human purposes are inevitably misguided by our own self-centered imaginations and value systems. As Gustafson explains, this does not free human agents from the responsibility to act, nor does it deny the possibility of accomplishing real good; it does, however, remind us of the ambiguity and limitation of all human moral action.[18]

Yet this perspective also, and perhaps more profoundly, leads to hope. Again, at the heart of Niebuhr's ethic of responsibility is the trust that the principle of being is not only powerful but also benevolent. With this trust comes "the liberty to interpret . . . all that happens as contained within an intention and total activity that includes death within the domain of life, that destroys only to re-establish and renew."[19] At the same time that it challenges our expectations of what is good and life giving, the theocentric view affirms that the power behind creation is indeed benevolent. With the humble relativization of human purposes comes hopeful confidence in the divine purpose.

There are examples of this combination of hope and humility among Christian responses to MTR. It is perhaps most apparent in AMERC, which seeks to establish mutual understanding and dialogue through the contextual education of church leaders. AMERC believes that this kind of careful listening can help overcome the assumption that the people of Appalachia are powerless and in need of rescuing; instead, it identifies "stories of hope" and positive, transformative visions that can lead to action.[20] AMERC's approach expresses humility, in that it seeks first to discern God's action and purpose in a particular context, but it also expresses a hopeful confidence that such action is already present and transforming the situation. In one example not directly related to MTR, AMERC director Lon Oliver points to a 2009 ABC news report that called attention to the crime and drug problems in Cumberland, Kentucky.[21] He argues that although this awareness was salutary, the report completely ignored the successful efforts by local churches and community groups to improve law enforcement and strengthen the relationship between the community and the police.[22] This kind of transformative power is often overlooked by typical discourses but highlighted by a theocentric view of power. A response to MTR that incorporates such a view would pay special attention to these kinds of grassroots movements.

Christians for the Mountains (CFTM), a group that has challenged MTR more directly than AMERC, believes that hope is one of the church's main contributions to this issue. Director Allen Johnson says that hope and joy are essential when confronting such an overwhelming issue—one that has driven many activists to despair. Without resorting to fatalism or what he calls "Pollyanna" optimism, Johnson sees Christian activists enacting their faith that God loves humans and the world and is the one power greater than evil.[23] Although it places less emphasis on humility and human limitation than AMERC, CFTM sees one of its main goals as nurturing this hopeful confidence in the benevolence of divine power. Practically speaking, this has involved providing pastoral support for those fighting against MTR to prevent the burnout and frustration that often accompany such an insurmountable task. Like AMERC, CFTM illustrates a hopeful awareness that divine power is at work in unexpected and transformative ways.

Identity and Inclusivity

Identity is enacted in dynamic and intertwined relationships and is itself dynamic and multifaceted; God the universal valuer relativizes all human conceptions of identity and boundaries. Niebuhr speaks at length about human identity in *The Responsible Self.* In Niebuhr's anthropology of person-as-answerer, the central characteristic of personhood is responding to a multiplicity of actions. The person-as-answerer is defined by her responses to other persons as well as to "that to which [others] respond," to the events and ideas of the world around her.[24] For Niebuhr, this relational anthropology raises the question of where unity—where identity—can be found in this network of relationships.[25] He contends that unity in the midst of this plurality is found only in the response of trust in "the radical action by which I am." One's unity as a self cannot be found in any of the conflicting and changing relationships in which one participates, nor in the finite causes to which one may devote oneself; rather, unity is found in the fundamental relationship, the existential action in light of which we interpret and respond to all other actions. Again, Niebuhr's value theory parallels this anthropology. Value is present wherever one being confronts another and these beings always confront one another in a state of becoming. Value is a characteristic of these relations between beings and between a being and its own process of becoming.[26] Just as one may turn in trust to the radical creative action to find unity in its manifold responses, the coherence of value in these manifold value relations is found in the center of value, which values all being universally.[27]

In bringing together Niebuhr's ideas about value and selfhood, I am not arguing that value and identity are the same thing for Niebuhr. Rather, relational value can be seen as part of identity: value, in Niebuhr's conception, is a fit between two selves, an expression of the worth or goodness of one self for another. Value can be positive or negative; fit can be good or bad. But in all the roles and relationships in which selfhood and identity are negotiated, value is present as one prominent aspect of those relationships. Indeed, to a large extent, value claims are implicit in identity claims: otherness is frequently used to suggest disvalue, while similarity implies value.[28] This is the dynamic behind the notion of a sacrifice zone: because a region is set apart as different, it can be dismissed as expendable.[29] Likewise, when the spokesman for Massey Energy asserts that protesters of a mountaintop mine are from out of state, the clear implication is that their voices are unimportant.[30] From a theocentric perspective, these claims represent value systems arbitrarily constructed around some finite starting point—in this case, a particular conception of identity.[31]

Identity, like value, is relational: rather than being a characteristic of a person or being in itself, it is present "wherever being confronts being." A person enacts identities in the roles and actions to which she responds, and these finite identities are no more monolithic or unified than those roles are. Moreover, Niebuhr argues that our own narrow self-images (that is, our conceptions of our individual and collective identities) are at the center of the destructive imaginations that divide us from one another. When these imaginations are reconceived with God, who values all being, at the center, we understand ourselves in continuity and unity with all of humanity and creation.[32] This unity is, from a theocentric perspective, the only absolute identity; other enactments of identity are partial and shifting, characteristic of the self who responds in a wide variety of interconnected relationships. These incomplete notions of identity are thus relativized—though not denied—by the revelation of God the universal valuer.

MOVING MOUNTAINS: SHIFTING IDENTITIES IN APPALACHIA

Based on this theocentric view of identity, claims of insider status and authentic Appalachian identity are, like conventional views of power, evil imaginations that center on human ideas and selfish interests. The previous chapter showed how a notion of Appalachian identity has been politically constructed to serve various interests. These constructions obscure the way people in the region construct a sense of themselves and their place "out of the flotsam and jetsam of many influences" in relationships with one another and with the rest of the nation.[33] In light of the above understanding of identity as enacted in relationship, and in light of a principle of being and center of value who relativizes all finite identities, this construction of Appalachian identity and demarcation of boundaries between insiders and outsiders, between who can and cannot represent Appalachia, reifies one finite conception of identity and establishes value judgments based on that human conception rather than on the inclusive intentions of the universal valuer. In Niebuhr's terms, these claims absolutize what is properly relative; recall that, for him, this is "the great source evil in life."[34] Moreover, notions of Appalachian identity are invariably constructed around the same issues: coal, poverty, nature, and history, for example. The label "coalfield" to describe certain areas of Appalachia exemplifies this kind of single-minded identity construction.[35] Niebuhr's anthropology reminds us, in contrast, that just as selves are constantly responding in relationship to other selves, they are also responding in relationship to other ideas and causes. The identities of Appalachians are forged in relation to a wide range of issues beyond coal or poverty; narratives of identity that limit themselves to these issues are incomplete.

Niebuhr argues that his anthropology of person-as-answerer is reinforced by the insights of other disciplines such as psychology, sociology, and history; likewise, this anthropology gives rise to a conception of identity that is similar to those emanating from other perspectives on Appalachia.[36] For example, Allen Batteau describes the invention of Appalachia as a process of renegotiating American identity and argues that the propagation of commodified images of Appalachian identity without attention to the relational contexts that created them inevitably falsifies these images.[37] Stephen Hanna believes that even an exploration like Batteau's leaves the dichotomy between insiders and outsiders intact and relies on the idea that there is some "real" Appalachian identity to which insiders have access. As I noted in the introduction, Hanna argues instead for the notion of intertextuality: that conceptions and narratives of Appalachian identity are mutually dependent and interrelated. Just as hegemony and resistance are intertwined, notions of insider and outsider construct and reinforce each other. Identity is produced around and across these categories; continued reliance on dualisms fails to recognize this.[38] Rebecca Scott believes that a conception of identity as shifting and fluid can be mobilized strategically to build coalitions for change. Rather than being concerned with the accuracy of claims about Appalachian authenticity, those involved in MTR can draw on a variety of identities and understandings in their struggle against "the divisive strategies of the industry, which support its unsustainable practices."[39] These perspectives share with Niebuhr the belief that identity is enacted in multiple dynamic and shifting ways across various relationships and that identity itself is relative and fluid.

IMPLICATIONS OF THE THEOCENTRIC VIEW: INCLUSIVE IDENTITIES

Scott's argument indicates the main implication of a dynamic and relative Appalachian identity in the response to MTR: a move toward greater inclusivity. If the radically inclusive center of being relativizes human boundaries and narrow conceptions of identity, and if revelation of this inclusive center means "that all our values are transvaluated by the activity of the universal valuer," the appropriate response is to conform our conceptions of identity and value to this radical inclusivity.[40] Gustafson's increasingly expansive relevant wholes exemplify this move (see chapter 3). Beyond Scott's point about coalition building, I want to consider inclusivity more broadly. In addition to the inclusion of people with a variety of different commitments, I believe this perspective commends a more inclusive theology (one that includes the perspectives of the more conservative rural congregations) and a more inclusive conversation (one that considers the larger relational context within which MTR is discussed). I begin, however, with the most straightforward aspect of greater inclusivity: breaking down barriers between different groups.

More Inclusive Coalitions Niebuhr argues that one function of revelation is to remove the self-centered imaginations that divide us from one another.[41] Scott echoes this idea by contending that a recognition of the fluid and shifting nature of identities can found new coalitions across traditional boundaries. For her, this means that opposition to MTR can bring together environmentalists and oppressed workers; it can mobilize Native American spiritualities along with Christian commitments and environmental ethics. Needless to say, Niebuhr has something more radical in mind. He envisions "human reunion" through "repentance and faith."[42] In the context of MTR, this requires challenging divisive claims of true Appalachian identity and the dualism between insiders and outsiders in order to engage in a truly inclusive dialogue. I do not imagine that such a dialogue will be easily achieved or that universally acceptable solutions will necessarily be forthcoming. The point is that efforts toward what Niebuhr calls "reunion" are an inescapable requirement of the monotheistic approach I am articulating.

Batteau argues that the dismantling of claims of Appalachian identity does not mean that conceptions of identity can no longer be employed for strategic purposes.[43] My approach also allows for this strategic mobilization. Certainly, the recognition that claims of identity are relational does not mean that they are necessarily illusory, any more than the view that value is relational means that it is purely subjective.[44] The relativization of all identity claims entailed by theocentrism, however, means that these claims must not be absolutized, reified into centers of value unto themselves. Yet awareness of the relativity and fluidity of all our understandings of identity, including our notions of Appalachia, can be mobilized in the service of unity rather than division. I believe that with this awareness, those involved on either side of the issue will be able to see that they have much in common; for example, many people on both sides hold similarly deep attachments to the mountains, even though they may value the mountains in a variety of (sometimes conflicting) ways.

Some Christians engaged in the debate see the value of such inclusive dialogue and are working toward it. The Reverend Stan Holmes, who worked as a coal supervisor before becoming an Episcopal priest, understands the coal industry. He, like many others, believes that if the conversation could focus on facts and a shared commitment that "we can do better," there might be real understanding between opponents and advocates of MTR. Unfortunately, he believes few coal operators would welcome such dialogue.[45] At the same time, Andrew Jordon, the owner of Pritchard Mining, regularly welcomes community and church members to his active mine site to show them the best and worst aspects of his operation. He, too, believes communication is key to exercising better and more responsible stewardship of the mountains.[46] Some fruits of this inclusive attitude are noted in the next section; here, the point is to show that the theocentric approach finds resonance with the attitudes of some Christians involved in this debate.

More Inclusive Theology   One boundary in this debate that receives little attention is the one between mainline, so-called county-seat churches, which are responsible for most environmental activism, and rural churches such as the Holiness and Pentecostal churches, which are more theologically and socially conservative, more focused on individual salvation, and less concerned with social issues like MTR.[47] Consistent application of Niebuhr's inclusive ethic entails the attempt to overcome these boundaries as well. Niebuhr is emphatic that division within the church is one of the most destructive effects of self-centered imaginations.[48] Moreover, I believe Niebuhr's monotheism may provide a more promising theological common ground than the principles of stewardship or social justice that frequently motivate members of mainline churches. The absolute sovereignty of God is a central commitment of Appalachian mountain religions, which are heavily influenced by Calvinism.[49] One Baptist minister's distinction between goodness and godliness illustrates the similarities with theocentrism. Arguing that much of what appears to be good in the world is deceptive, he exhorts, "A Christian must never ask, 'Is it good?' He must ask, 'Is it from God?'"[50] This affirmation, which is largely representative of mountain religion, has affinities with an approach that places God at the center of all moral interpretations and value systems. Based on these common theological commitments, it is plausible that an approach to MTR based on God as the principle of all being and the center of all value may allow greater engagement with these rural traditions.

Furthermore, as with other human boundaries, the distance between mainline churches and mountain churches is not as great as might be imagined. "Indigenous" mountain traditions have had a great influence on the overall religious character of the region, including on county-seat churches.[51] Even mainline environmental activism exhibits a typically Appalachian emphasis on personal transformation and conversion.[52] And Holiness and Pentecostal traditions have been active in struggles for unionization and economic justice in Appalachia in the past.[53] In this context, an inclusive theocentrism may make greater cooperation between county-seat and rural churches possible; in any case, it is my contention that it makes such cooperation necessary.

A More Inclusive Conversation   Finally, and perhaps most importantly, greater inclusivity requires an expansion of the discussion of MTR to include the larger context that shapes it. The various political constructions of Appalachian identity addressed in the previous chapter, particularly Rebecca Scott's, illustrate how conceptions of that identity have been developed and promulgated in tandem with narratives about coal. Parts of Appalachia are constructed as coalfield; the dirty, physical, abject nature of mining further establishes Appalachians as other, and that otherness, in turn, reinforces the belief that the rest of the nation is effectively isolated from mining and from the environmental and public health and safety problems that go with it. Discussion of MTR as an isolated issue, without consideration of the various contexts of which it is a part, only perpetuates the idea that Appalachia, Appalachians, and Appalachian issues are different—a problem to be solved, to be sure, but always at a safe distance.

Thus, the theocentric challenge to reified identities demands that the church expand the conversation to include the contexts in which MTR is understood. Niebuhr's relational anthropology also supports this move, since it argues that selves exist in relationship not only to other people but also to a "third thing," to the causes or ideas to which those others respond. Responding to another person's action also means responding to what she responds to, to her own interpretation of the world and her own cause.[54] In addressing MTR, this means interpreting and responding to the issue and to the actors involved in it, in light of the other causes and relationships that shape their interpretations and responses.

To some extent, this study is an attempt to do just that: to understand MTR not as an isolated issue but as one aspect of a history of struggles to imagine Appalachia in order to reshape these imaginations into more theocentric ones. This is one way of responding to the issue's larger context. But there are other contexts that need to be addressed in this conversation, by both sides. Both opponents and advocates of MTR need to see it in the context of the nation's energy policy, which is shifting away from coal but, many would argue, cannot yet escape its dependence on coal. Much of the nation currently relies on energy from coal (although, as opponents are quick to point out, little of this is from MTR operations), and some discussions acknowledge this dependence. But unless the debate over MTR and mining includes a discussion of national energy policy, this debate will be limited and shortsighted.

Another related aspect is the broader economic context, especially in places where the economy is centered on coal (even though this dependence is sometimes exaggerated by advocates of mining). Even as many opponents

of MTR argue that the economic impact of ending the practice would be minimal, they recognize the need to articulate alternative visions of the region's
future economy.[55] At the same time, as Scott shows, many Appalachians are
unable to envision work as anything other than mining.[56] These claims are,
of course, contestable, which is exactly the point: they need to be included
in the conversation, particularly as it is represented to a national audience,
rather than presupposed. Statements like the denominational resolutions
discussed in the previous chapter, for example, address MTR in isolation,
treating it as an issue for one particularly problematic region of the country,
which only perpetuates inadequate imaginations of Appalachian identity.

Examples of Inclusive Identities Christians on both sides of the debate are
challenging reified conceptions of Appalachian identity and making efforts
to achieve greater inclusivity. Again, the contextual educational approach
of AMERC is illustrative. By encouraging church leaders living outside the
region to explore theological questions in local communities, AMERC shows
students how the identities of Appalachia and Appalachians are enacted in
relationships with other people and in relation to issues other than those
typically associated with the region. In AMERC's classes, Appalachia is
understood not only in relation to poverty or mining, nor is poverty understood only in relation to Appalachia; rather, these issues are explored in
relation to, for example, prophecy or eschatology. AMERC focuses on con-
necting its students with mountain churches such as Mountain Pentecostals
and Old Regular Baptists. Thus, the complex network of relationships that
shape fluid identities is engaged. Moreover, AMERC's work is based on the
belief that "the Gospel is borne on local culture" and that the purposes and
values of God are more inclusive than the human boundaries that separate
one group from another. Director Lon Oliver has seen numerous instances
of this openness leading to "life-giving discussion," such as a staunch environmentalist who admitted, in the spirit of real dialogue, that a reclaimed
mine site was indistinguishable from the untouched mountains around it.
Oliver also cites similar admissions from MTR supporters that some mining
and reclamation activities have been truly irresponsible.[57] These real changes
in attitude are the fruit of AMERC's inclusive approach.

As noted earlier, mine owner Andrew Jordon regularly invites members
of his church community to his MTR site and even holds Bible studies there.
He seeks to expand the conversation to include energy policy and Appalachia's future, even as he narrows the focus to one particular mine (small

in MTR terms) and its impact on the community and individual workers' lives.[58] Efforts toward greater inclusion are not limited to opponents of MTR.

Meanwhile, leaders like Allen Johnson of CFTM point to hopeful stories of genuine understanding between activists and miners. At a prayer service on an active mine site in 2008, participants engaged in real, positive dialogue with the miners who had gathered to disrupt the service. The miners articulated their own conflicted position—concern about the effects of mining versus the need to support their families in a place with few other options.[59] From a theocentric perspective in which identity is fluid and relativized by God the universal valuer, inclusive efforts like these are to be celebrated, and the divisive claims and imaginations that foreclose on them must be challenged and reevaluated.

Reclamation for God

Human changes to the environment, including destruction and reclamation, are part of God's action in the world and can reflect divine purposes to a greater or lesser degree. Articulating a theocentric view of destruction and reclamation is more challenging than doing so for power and identity. Obviously, Niebuhr did not address mining or any other environmental issue at any length. Nonetheless, just as the statements about power and identity represent positions derived from and consistent with Niebuhr's monotheism (rather than positions he explicitly advocated), I believe the above statement reflects a consistently Niebuhrian theocentric position on destruction and reclamation.

Niebuhr's straightforward formulation of the ethic of responsibility affirms, "God is acting in all actions upon you. So respond to all actions upon you as to respond to his action."[60] Clearly, this entails interpreting human intervention in the environment, including destruction (whatever that might mean) and reclamation, as actions of God and responding to them as such. Yet in response to the challenge that this leads only to resignation and fatalism, Niebuhr again raises the myths or imaginations we use to interpret action and formulate our response. If our imaginations are those of resignation or indifference, then they are not imaginations of the divine creative power who interacts with creation. The God who acts in the world is not a pure determining power, the "foreordainer of all that happens"; rather, this is the loving, dynamic God revealed in history and in Jesus Christ.[61] It is the principle of being that both limits and affirms all finite beings. Response to this God requires interpretation of God's actions, based on experience and revelation, and discernment of which responses are most in line with what can be understood of God's purposes, which are emphatically not to be equated with human purposes.[62]

Human agents inevitably interact with the world of nature and are therefore participants in God's action in the world. At the same time, according to Gustafson, the theocentric view demands that they seek to understand the ultimate purposes of that action even as they participate in it.[63] Because they can and necessarily do interact with nature, humans are accountable for the choices they make, yet these choices are always ambiguous. Harmony and equilibrium, if they exist, are dynamic and imperfect.[64] As Gustafson argues, "there is no clear overriding telos, or end, which unambiguously orders the priorities of nature and human participation in it so that one has a perfect moral justification for all human interventions."[65] Nonetheless, he insists that because we are accountable for our interventions, we must continually seek to understand the divine purpose and, to the extent possible, conform our actions to it.

RECLAMATION RECONSIDERED

It is in this context that destruction and reclamation, like any human intervention, must be understood. As part of human participation in God's purposive action in the world, they are inevitably ambiguous. Without a clear knowledge of God's telos, we are forced to admit that there is no absolute justification for any particular moral stance in this case. There are interrelated values, benefits, and consequences to be considered in light of the various choices described in the previous chapter: choices about the original state of the land, how human impact is measured, and what the appropriate use of a place might mean. As I argued there, we are better off thinking of these terms as descriptors of how humans change places; Niebuhr's and Gustafson's language of ambiguous participation reinforces this. If, as the latter contends, this ambiguity of human interventions leads us to recognize that "God is in the details" or that the morality of our actions lies in their specific consequences, then our discussion of how we change places will need to be more precise than terms such as destruction and reclamation allow.[66]

Accordingly, the assumptions implicit in Christians' use of these terms can be seen as representing, from my theocentric perspective, inadequate imaginations. When MTR is approached with the a priori belief that it destroys something God created and values, this presumes to know what

God values and how, and it presumes that God's valuing is static and singular as opposed to dynamic and universal. That God values mountains is certain, since Niebuhr affirms that God values being universally; how God values mountains, a particular mountain, or a particular part of a particular mountain is a question that must be approached with much greater humility if we are to avoid absolutizing the relative. Likewise, the a priori assertion that because only God creates mountains, reclamation must be blasphemous denies God's ongoing participation in creation through human (as well as nonhuman) agency and forecloses the possibility of discerning how to make reclamation more closely reflect the divine purpose.

The theocentric perspective therefore commends greater attention to the details—the values and disvalues of human interaction with the environment, and the choices and relationships in which those values are negotiated—because these details allow the purposes of God to be discerned, so far as that is possible. There are better and worse ways of interacting with nature; interventions can be more or less conducive to God's intentions. As with notions of identity, it may be that discussions of destruction and reclamation can be strategically mobilized in effective and nonabsolutizing ways; yet it is important to bring to light the choices and assumptions behind them. Even those who are firmly opposed to MTR may recognize the importance of substantive discussion about what kinds of extraction and reclamation are being practiced on mines that already exist. When, in the midst of disagreement on adequate standards for reclamation, Andrew Jordon takes over abandoned and previously unreclaimed surface mines for the purpose of mining and reclaiming them, perhaps even opponents of the practice might see some trace of God's sustaining actions in Jordon's work.[67]

IMPLICATIONS OF THE THEOCENTRIC VIEW: RECLAIMING THE MOUNTAINS FOR GOD

A theocentric perspective on reclamation suggests a move toward a more substantive, precise discussion of the changes to the environment we believe to be warranted or unwarranted, in light of our limited knowledge of divine purposes. Jordon provides one example of this kind of engagement. There is certainly room for disagreement with his claim that he is practicing good environmental stewardship by carefully attending to the reclamation of his mines, yet this disagreement may lead both sides to consider what constitutes effective reclamation. Even if agreement is impossible, a more complete understanding may be attainable. AMERC, again, provides a compelling example. Even as Oliver argues, referring to MTR, that "some things need to be stopped," he believes it is crucial for Christian activists, particularly from mainline churches, to pay more attention to the real work being done toward effective reclamation, as this reflects part of "what God is up to" in this context.[68]

The work of systems ecologist Samir Doshi is especially important in this respect. Doshi argues that the effort to improve reclamation does not implicitly condone further mining; rather, he insists, it "reflects a commitment to ensure the survival and sustainability of communities that have been marginalized."[69] He points to projects at Virginia Tech and other institutions that result in greater tree-reestablishment rates at a lower cost than current techniques. Efforts like Doshi's (and Oliver's and Jordon's) to engage in thoughtful and thorough discussion about people's impact on the mountains are precisely what a theocentric approach requires.

Another related approach to this issue can be derived from the theocentric view. If human interventions are interpreted as part of God's action in the world, and if the goal of the theocentric ethic is to direct those interventions toward what we believe to be God's purposes, I propose that the notion of reclamation might be broadened. We might think of it in terms of reclamation on God's behalf—that is, reclaiming mountains, and humans' relationships with them, for God's purposes. This would include viewing conversations about values and disvalues in the reclamation process as one way of participating in God's sustenance of creation, as described above. Whatever one's convictions about the practice of MTR, there are sites waiting to be restored, and responsible reclamation of the sort Jordon claims to practice can, I believe, be understood as participation in God's action. Even in the context of dynamic and multiple value relationships, few people would disagree that a responsibly reclaimed mine is more valuable than an unreclaimed one.

My idea of reclamation for God, however, can be broadened even further, beyond the practices typically thought of as reclamation. It can include the symbolic practices Christians employ to claim a mountain or part of a mountain as belonging to God, as part of God's good creation before, during, and after mining activity. Jordon's mine-site Bible studies are one example, expressing the belief that God can be encountered anywhere—even in the midst of an active mine site. The liturgies practiced by groups such as CFTM and the Catholic Committee of Appalachia are other powerful examples of reclaiming the mountains for God. In these services, Christians pray for

the health and renewal of the mountains and of the mountain communities around them, and they mourn the greed and thoughtlessness that
are destroying that environment for the sake of cheap energy. In one such
service, worshippers carefully planted wildflower seeds on a mined mountaintop, powerfully symbolizing hope and faith in God's creative and sustaining power in the most damaged of places.[70] Another intriguing example
is the partially successful effort by the Ohio Valley Environmental Coalition and the West Virginia Council of Churches, in collaboration with the
West Virginia Coal Association (normally an opponent), to strengthen the
legal protections for cemeteries and families' access to them. Recognition
of the sanctity of these small plots provided common ground (literally) for
both sides to acknowledge and defend the limits of human intervention in
God's creation.[71]

These practices, though very different, all powerfully illustrate the divine
creative action that continually upholds and embraces creation, both through
and outside of human activity. They symbolically reclaim the mountains for
God. This theocentric imagination of reclamation can interpret physical restoration of ecosystems and symbolic acts of reclamation as serving a common cause, reminding us of the ultimate, original source of the mountains'
value, and participating in the sustaining actions of that source.

In this chapter I have described consistently theocentric perspectives on
the discourses explored in detail in chapter 4. In other words, I have suggested more God-centered imaginations in place of inadequate or destructive imaginations. I have outlined a view of power as belonging ultimately
to God, such that any power exercised in creation is seen as an expression of
divine power. This perspective can inform hopeful practices that call attention to the unexpected and transformative ways divine power is at work in
creation. I have articulated a view of identity as fluid, dynamic, and relative,
in contrast to Niebuhr's concept of God as the universal valuer, who breaks
down all finite boundaries and self-images. Based on this view of identity, I
have described more inclusive practices in terms of broader coalitions, more
comprehensive theological commitments, and greater attention to the larger
contexts that inform the MTR debate. Finally, I have reconsidered destruction and reclamation in light of the belief that God acts in and through all
creaturely actions. This belief leads to the conclusion that destruction and
reclamation, like any other human intervention, are part of God's action and
that humans, as interactive participants, are accountable for conforming to

God's divine purpose (to the best of our ability). Accordingly, I have argued for more substantive discussion of the interrelated values and disvalues of mining and reclamation and for a conception of reclamation that includes all the practices mobilized by Christians to reclaim the mountains for God symbolically as well as physically. The discourses addressed here are only a few of the various narratives and ideas that shape our understanding of MTR, and the recommendations I offer are only some of the possibilities for responsible action. One of the most salutary aspects of a theocentric approach is its insistence that moral action, as a response to the actions and responses related to a given situation, can finally be addressed only in particular contexts.

In chapter 3 I responded to the charge that Niebuhr's ethic is too conciliatory, that it undermines a strong prophetic challenge to a practice like MTR. This concern arises again at this point (and comes up once more in the final chapter).[72] The theocentric emphasis on divine agency and the consequent downplaying of human agency may seem to entail passive resignation or the simple affirmation of all actions as "part of God's plan." The perspectives I have outlined and the recommendations I have offered tend to involve increased dialogue and understanding, as opposed to dramatic actions and public outcry. Indeed, some of the activists cited in this chapter might be disappointed to see their views marshaled in support of a seemingly overly irenic approach. And as I conceded earlier, to some extent, the theocentric approach, with its characteristic humility, nuance, and self-examination, must necessarily shy away from strong oppositional stances. Yet, if these more oppositional stances rely on destructive imaginations, if they absolutize the relative by placing human values or interests at the center of their worldviews, then, from the theocentric perspective, they ought to be eschewed.

This criticism of Niebuhr's theocentric ethic misunderstands its founding impulse. Niebuhr's ethic is consistently and insistently focused on the purposes of God, a focus that is radically different from the multiple finite centers of worldly value systems. Even as the actions and situations to which agents respond are the same, the theocentric interpretive framework radically changes the character of those responses; it grounds them in a center that is not only absolute and unified but also revealed as wholly different from all other value centers. As Niebuhr argues, the symbols of fatalism are simply not the right symbols: the deistic "Determiner of Destiny" is not the loving and transforming God and Father of Jesus Christ.[73] Most significantly, the latter God is one whose power is expressed paradigmatically in apparent weakness. In this light, what seem to be conciliatory and even placatory responses can be interpreted instead as humble and hopeful expressions of trust in the power of the creator and the valuer of all being.

Nevertheless, there is a temptation to retreat into the Niebuhrian attitude of humility and confidence in God's agency and thus postpone responsible action indefinitely in the name of ever-elusive understanding. Such an attitude seriously misconstrues this approach. It ignores Niebuhr's contention that we not only are acted upon and limited by others but also act upon and limit them—that God's interaction with the world works not only on us but also *through* us. Put another way, in Gustafson's terms, this attitude of quiescence focuses solely on the ambiguity of our interactions with our world, to the exclusion of the corollary recognition of our accountability. Discernment is needed to resolve this tension between self-critical, humble understanding and responsible action, however ambiguous. In the next chapter, I offer some constructive suggestions for this process of discernment. For the present, suffice it to say that for all its emphasis on humility and ambiguity, a theocentric approach begins and ends with action: the person-as-answerer necessarily responds and is accountable for her or his response. Consistently applied, this approach brooks no endless vacillation.

A related criticism from the opposite side is that this approach claims too much: my appeals to the possibility of real understanding may seem virtuous and reassuring, but they are unrealistic; the stakes are too high, and the two sides cannot find common ground. I believe this is empirically and theologically false. Empirically speaking, in the MTR debate, as in past struggles over coal, there are examples of understanding between representatives of the different perspectives. Efforts to achieve more open, more inclusive dialogue have borne real fruit, bringing fierce opponents closer to mutual understanding. Several examples have been noted: a staunch environmentalist who recognizes the reality of good reclamation, Jordon's open invitation to anyone who wants to see all facets of his operation, the collaboration of erstwhile opponents to protect cemeteries, the sharing of concerns between worshippers and the miners who originally intended to disrupt their service. Moments like these point to the possibility of real understanding. Additionally, the criticism is theologically inadequate from the theocentric perspective of confidence in the divine power. Again, at the center of Niebuhr's theology is trust in a benevolent being acting within creation. Approached with this confidence, the apparent success or fail-ure of dialogue is not the main issue; for individuals who believe God is at work in those who limit us as well as those who support us, any attempt at greater understanding of others represents an advance in understanding God's purposes in creation.

6

Loving the Mountains

Conclusions, Challenges, and Ways Forward

Up to this point, I have articulated and applied a theocentric approach to MTR. This approach has involved a critical examination of the discourses surrounding the debate in a search for more God-centered imaginations of power, identity, and reclamation. This process of examination, however, does not represent the whole of a theocentric ethic. Since the foundation of Niebuhr's ethic is a relational anthropology in which all action has the character of response, moral action is never isolated; rather, it is always part of a whole life, lived in the context of numerous relationships. The theocentric transvaluation that Niebuhr describes is not, in the end, simply the reexamination of a select set of values or imaginations; rather, it is *metanoia*, the creation of a new self and a new community.[1] Niebuhr's work resembles virtue ethics in this respect: ethical decisions take place in the context of an entire life within a community and therefore must be reflected on and discussed in that larger context.

In describing a theocentric project of critical examination and rehabilitation of imaginations around MTR, this study has focused on one moment in the life of the new community. I have chosen this moment because it is both a pivotal aspect of the overall theocentric ethic and a particularly distinctive feature of this approach. In a sense, this moment establishes the conditions for the possibility of theocentric action. Nonetheless, it would be inconsistent with Niebuhr's thought to treat this central part of his ethic in isolation, as if it were the whole. Thus, in this chapter I place the process of critical examination and transformation of imaginations in the larger context of the moral life of the church. I describe this process as a central feature of a larger ethic of theocentric moral action that includes relationships not only with God and other persons and communities but also with particular places.

I begin by summarizing the project and bringing together the multiple strands discussed thus far in order to survey this approach as one coherent whole. Next, I integrate this approach within the life of the church, describing what I call a pattern for theocentric moral action: a set of concrete practices intended to enact and sustain the theocentric imaginations I have already described. I then return to a feature of theocentric moral action that Niebuhr does not explicitly address: its relationship to place. I do this by responding again to the criticism that this approach, like Niebuhr's ethic in general, is too passive or conciliatory, that it leads only to resignation and undermines any possibility of a strong prophetic challenge. I argue, first, that this criticism misunderstands key features of Niebuhr's position and represents the absolutizing of the relative that he warns against. Second, I consider how my theocentric approach to MTR might respond to this criticism through careful and active attention to the places of Appalachia—that is, by really loving the mountains. Finally, I argue that viewing MTR as representative of a new geological epoch—the Anthropocene—grounds what may be the strongest theocentric challenge to the practice.

As I have argued throughout, the approach I propose focuses far more on attitudes and practices than on moral judgments about particular decisions or actions. This is consistent with Niebuhr's suspicion of any ethic that implicitly makes our understanding of God conform to human values and interests, rather than vice versa; to base ethical reasoning on certain values, abstracted from concrete relationships among persons and between persons and God, is to absolutize the relative. Nonetheless, a theocentric approach can and should provide some guidance for moral action, and at the end of chapter I articulate what that guidance might be. These guidelines suggest the general direction indicated by my approach, some plausible outcomes of the process of examination and criticism and the pattern for moral action I describe.

The Theocentric Approach to Mountaintop Removal

I began this study, after a brief introduction to the practice of MTR and some of the problems and questions it raises, by considering research on the far-reaching environmental, economic, and political effects of MTR. Arguing that the interpretation of these data is shaped by values, I turned to a variety of environmental ethical perspectives that respond to the social complexity of environmental issues. What these approaches have in common is a recognition that the conceptual frameworks and values invoked in discussions of environmental issues are the product of intricate social processes, and that an adequate response to MTR must attend to this social construction of value. At that point I turned to Niebuhr's conception of value as relational, "present wherever being confronts being, wherever there is becoming in the midst of plural, interdependent, and interacting existences. It is not a function of being as such but of being in relation to being."[2] Niebuhr argues that for a system of value to have coherence, it must have a center—the being or beings in relation to whom whatever is good is good. He affirms that only a system organized around a transcendent, absolute, universal center of value—that is, God— is capable of recognizing and responding to the fundamental integrity and divinity of all creation; other value systems only contribute to the divisiveness and confusion of human agency. For this reason, a Niebuhrian theocentric ethical approach, based on this relational understanding of value, is ideally suited to what geographer Stephen Hanna calls the intertextuality of Appalachia. Rather than challenging certain narratives, images, and stereotypes of Appalachian identity in light of some supposedly real idea of Appalachia, the theocentric approach allows these imbricated and mutually dependent texts to speak while relativizing them in light of the transcendent center of value.

Turning from Niebuhr's value theory to the anthropology, theology, and morality it supports, I articulated in chapter 3 the basics of a theocentric approach to MTR. I described Niebuhr's anthropology of person-as-answerer, whose interpretations and imaginations shape responses to actions, and his theology of God-as-valuer, whose universal purposes ought to stand at the center of all human moral imaginations. These two elements lead to a view of moral action that calls for a critical examination of all finite interpretations and systems of valuing and the displacement of selfish and human-centered imaginations with theocentric ones. The church in particular must engage in this critical self-examination.[3] The God who is the central value and the fundamental actor to which we respond does not necessarily, Niebuhr emphasizes, value what we value; whatever may or may not be discerned about the values of the divine valuer, it is clear that they must be more inclusive and less limited than our finite conceptions of value. Our systems of value and moral imaginations are thus inadequate to the extent they center on human values rather than seeking divine ones.

Conforming our values to those of God the universal valuer, therefore, means eschewing the finite imaginations that, in Niebuhr's words, absolutize the relative, placing some finite, usually selfish, value at the center of our interpretations rather than the inclusive, universal purposes of God. Theocentric imaginations, because they view all being as valuable in relation to God and all actions as reflecting the actions of God, can find coherence in the midst of complexity and multiplicity, rather than denying complexity and imposing artificial unity or conformity. As noted in chapter 3, the work of James Gustafson provides a rich example of how to attend to this complexity in an effective and discerning way, through interdisciplinary inquiry and more expansive relevant moral wholes; Gustafson also expresses the appropriate attitude of humility in dependence on God's action. The work of Emilie Townes is helpful as well, particularly her use of counterhegemony and its tool countermemory. Counterhegemony necessarily threatens hegemony not as one truth in opposition to another but as multiple truths in opposition to hegemony's pretense of exclusivity. Thus, a theocentric approach that replaces finite and absolutizing imaginations with inclusive and multivocal theocentric imaginations is a project of counterhegemony. If God is at work in all aspects of creation, as Niebuhr's theology affirms, then, as Gustafson argues, "God is in the details."[4] In this case, no single monolithic (or hegemonic) worldview can encompass the divine purpose.

After exploring this theocentric approach in the work of Gustafson and Townes, I applied it to MTR in subsequent chapters, examining dominant narratives of power, identity, and reclamation and describing more inclusive theocentric imaginations of these concepts. In chapter 4 I traced the way these narratives have been constructed and the perspectives and voices they neglect, arguing that they constitute what Niebuhr calls destructive, inadequate, or evil imaginations because they are centered on finite and divisive interpretations rather than inclusive ones. I then indicated how these narratives have influenced many Christians' and Christian groups' responses to MTR. I did this not to criticize or refute their positions but to engage in the critical self-examination Niebuhr calls for, as part of the rehabilitation of imaginations in light of God's purposes and values. The church's approach to the freighted issue of MTR cannot rely on simple binaries and dualistic narratives that divide people—binaries such as power and powerlessness, insiders and outsiders, and destruction and reclamation. Instead, a theocentric approach demands inclusive imaginations that seek the more complicated realities of the mountains. Rather than imposing clarity and unity by denying complexity and interrelatedness, theocentric imaginations find unity in the source of being and the center of value around which they revolve. For this reason, these imaginations rely on Townes's countermemory—voices and stories that are missing from the dominant narratives, such as examples of power in unexpected places, cooperation that breaches the insider-outside divide, and the many ways values are negotiated in practices of reclamation, whether physical or symbolic (or both).[5]

In chapters 4 and 5 I addressed the contrasting imaginations of the three concept-pairs—power-powerlessness, insiders-outsiders, destruction-reclamation—separately, considering the limitations of each of the dominant narratives and describing what more theocentric imaginations would look like in each case. Yet as Niebuhr's moral anthropology insists, and as the examples provided in chapter 5 affirm, a theocentric worldview is ultimately unitive: because all moral action has the character of response to other persons and values, no single aspect of an issue can be abstracted from its other aspects. For example, a more inclusive imagination that seeks to undermine the insider-outsider boundary—the us versus them dynamic—would inevitably lead to greater attention to the unexpected workings of power outside its usual loci; reclaiming for God is just one way of seeking a more inclusive discussion of the issue. The examples of AMERC and mine owner Andrew Jordon, among others, are indicative of an approach that is more inclusive and, in that respect, theocentric in multiple interconnected ways.

Moreover, as indicated in chapter 5, a theocentric approach to MTR must be unitive not only in the sense of dealing with the practice holistically but also in the sense of considering it against a whole array of related issues: economic alternatives, energy policy, global warming, sexism, racism, and classism, to name a few. All value, all being, is relational, and the divine purposes that underlie theocentric imaginations can be known (to the extent they can be known at all) only by attending to the rich intertextuality of these relationships. God's will with regard to a particular issue can be discerned only in relation to a range of other issues. Again, the temptation to engage in vacillation and unending circumspection looms, made even more tempting by the Niebuhrian attitudes of humility and dependence. The challenge, therefore, is to engage in this process of critical examination and discernment of the divine purpose in order to act; for ultimately, God acts not only upon us but also through us.

Revelation as Mountain: A Pattern for Theocentric Moral Action

Thus far, the challenge of putting theocentric imaginations into action in accordance with God's purposes has been discussed primarily in general terms. Some characteristics of these divine purposes have been identified;

most fundamentally, the purposes of God as the universal valuer must necessarily be universal and inclusive. Accordingly, these purposes are opposed to imaginations that sow conflict and lead to conflicted agencies and divided loyalties, and the church ought to oppose such imaginations as well. Beyond these somewhat vague suggestions, Niebuhr yields few clues to how the church might know and act on the divine purpose in a given situation. This is consistent with his insistence that moral action is always relational and, for that reason, irreducibly contextual. Ethical decisions are not made in the abstract; the divine purpose is known only in particular situations. Nonetheless, if my theocentric approach is to be meaningful in more than a theoretical way—that is, if it is to be useful to anyone—some concrete practices need to be explicated. In this section, rather than articulating specific theocentric ethical imperatives, I consider theocentric action as part of the whole life of the church. I describe what I call a pattern for theocentric moral action, identifying a set of practices through which the critical examination and transformation of imaginations might be enacted in the church's common life. To make these practices concrete, I offer examples where applicable. I draw here on both Gustafson and Niebuhr in a general way, but the delineation of these practices is my own.

Some of these ideas have been touched on in previous chapters, and the suggestions here build on those earlier explorations. The critical examination of imaginations is a key part of an ethical response to MTR, but as I noted earlier in this chapter, this self-examination and self-criticism make sense only as the self-reflection of a living, acting agent in the context of a community.[6] The goal here is to integrate this kind of reflection into the common life of the church, since, in Gustafson's words, "if the Church is called to accept responsibility for society and culture it must develop in thought and action the most pertinent and effective ways of fulfilling this vocation."[7] I believe that these practices, like Niebuhr's approach as a whole, are applicable to the moral life of the church in general, not just to MTR or similar issues. But because this study is focused on MTR in Appalachia, I draw examples from that specific issue and region.

I make these suggestions for the church, rather than for individual agents, for the same reason that Niebuhr does. Moral action is always necessarily relational; we act, respond, and interpret in relationship with others. Certainly, we do this in a wide variety of communities, and the church is not likely to be the most (or even one of the most) influential.[8] Nonetheless, as Gustafson argues, the church is the community in which one knows oneself to be responsible to God, accountable to God for one's actions, and valued by God.[9] Thus, it is to the church community that this pattern for moral action is directed.

THE CALL TO REPENTANCE

An attitude of humility is a primary product of Niebuhr's ethic (see chapter 3). The dependence on and trust in "the radical action by which I am," which is fundamental to the ethic of responsibility; the relativization of all finite centers of value and partial interpretations; and the insistence on the priority of divine action behind all creaturely actions all necessarily lead to the humble recognition of the limitations of human moral action. As Gustafson articulates most clearly, a theocentric approach involves the awareness that value is always relational and that value and disvalue often coexist; that is, what is a value in the context of one relationship may represent a disvalue in a different relationship.[10] Moral choices therefore frequently involve the sacrifice of some value for the sake of another. The theocentric approach also includes the affirmation that humans are accountable to God even though they can discern God's purposes only partially.[11] Wherever else such an approach to ethics leads, humility and awareness of moral ambiguity are its chief characteristics.

Moreover, revelation—that is, the particular revelation in Jesus Christ that illuminates and transforms all aspects of the life and history of the church—leads the church to view itself and the world as God sees them, in light of God's inclusive purposes. As a result of this revelation, the church can only lament its actions to divide persons from one another rather than unite them, and its placement of itself—or its nation or causes—rather than its God at the center of its imaginations.[12] Repentance of its own participation in destructive imaginations is the beginning of the church's effort to rehabilitate its imaginations and make its moral responses more theocentric. This kind of repentance is addressed later with respect to the theocentric condemnation of the Anthropocene epoch.

Yet Niebuhr is insistent that "the problem of human reunion [that is, the establishment of more universal, unifying imaginations] is greater than the problem of church reunion" and that the former, like the latter, can be accomplished only through repentance and faith.[13] In the Bible, the prophetic voice first calls for repentance; the church must continue the prophetic duty of preaching repentance to society, even as it affirms the good news of God's mercy.[14] As a community formed and transformed by revelation, the church should remind everyone of their complicity in destructive imaginations, beginning with itself: "In ethics [the church] is the first to repent for the sins of a society and it repents on behalf of all."[15]

Allen Johnson, director of Christians for the Mountains, exemplifies the call for repentance in his reflection on Christian responses to MTR. He argues that attention to the "deep truth" of what God is seeking to do necessarily leads to repentance in the face of our own complicity.[16] With respect to MTR, this complicity is readily visible: households all over the nation use electricity derived from MTR. It is also present in the pervasive assumptions and stereotypes about Appalachia that continue to alienate groups from one another. In this and other issues, however, the call for repentance is only the beginning of responsibility; as Gustafson insists, even as we are reminded of the inevitable ambiguity of our actions, our accountability to act as participants in God's order is inescapable.

CONVERSATION AND LISTENING

Gustafson describes the ethical implications of *The Responsible Self* as follows: "The intellectual work of ethics, like the moral life described in this book, always takes place in community; it is a dialogue with others present to the thinking [person]. Thus communication with others is a part of the work of ethics itself. It is in this process of communication that the understanding of the world in its moral nature has its effect upon the moral outlook and actions of particular persons."[17] The relational nature of all moral action and the belief that all the actions and events of creation express God's purposes require the responsible self to be as open as possible to many different perspectives. Yet what Gustafson is describing, and what I am advocating, is not merely an attitude of open-mindedness or communication in the abstract; it is the actual practice of engaging in conversation with and (more important) listening to others with different points of view. Niebuhr affirms that "external" views and histories—how the church is seen by those outside it—are as integral to revelation as the church's own "internal" beliefs about itself and its world, and they are uniquely valuable for the community's self-examination.[18] And as Gustafson explains, the primary way God acts upon us is through those who act to limit us.[19] The only way to access these external views and understand these limiting actions is to pay attention to others.

Examples of this kind of real communication have already been described, and once again, AMERC is particularly illustrative. An emphasis on listening pervades all of AMERC's work, which seeks, in Niebuhrian fashion, to understand "what God is up to" in a given situation.[20] This approach has led to opportunities for a real exchange of opposing views. Many others in the region emphasize the need for more dialogue and better mutual understanding, particularly with the miners themselves. And there have been moments of real understanding between individuals on different sides of this issue. Even without such dramatic successes, however, if God acts in those who limit us, such dialogue is necessary to discern God's purposes.

Certainly, neither the church's internal views nor the external views presented by others are revelatory in themselves. Revelation is ultimately God's own disclosure of divine purposes.[21] The church's open-minded listening, therefore, does not affirm all views equally; rather, it interprets them in light of what is known and can be discerned of God's action in the world. This interpretation draws on the interdisciplinary exploration Gustafson utilizes, as well as the critical examination I have been articulating, to seek out the divine purposes within related and conflicting perspectives (discussed in the next section). Yet if all creaturely actions are expressive of God's action, there can be no substitute for real conversation and close listening in the search for clearer knowledge of God's purposes.

CRITICAL EXAMINATION

Discerning God's purposes in a given situation requires not only attentive listening to other perspectives but also critical examination of both those perspectives and our own. As Gustafson explains, "effective moral action depends upon astute and accurate analysis of the social system in general and of the problem situation in particular."[22] All history is expressive of God's action, but discerning how the divine purpose might be revealed in a given event is a matter of ongoing reflection and interpretation.[23] I have already explained at some length how this examination is related to the project of establishing theocentric imaginations focused on God's purposes. Here, I argue that this process of analysis must be an ongoing feature of the church's moral life.

In particular, a close critical examination requires a special awareness of and attention to the operations of social power.[24] As Townes confirms, the multivocality and complexity demanded by the theocentric imagination are opposed to the hegemonic discourses that attempt to impose unity through exclusivity. It is toward these discourses, and the dynamics of power that support them, that a critical eye must be turned. In addition, understanding social power is essential because the radical transformation of values that results from revelation is most notably a revolution in our ideas about where and how power is exercised.[25] These aspects of the theocentric imagination call for a close examination and criticism of the workings of social power.

To be clear, I am suggesting that the church create opportunities to talk, both among church insiders and with those outside it, about the kinds of narratives described in this book and the ways they are or are not reflective of God's purposes. One Appalachian organization that exemplifies this kind of analysis, albeit not from an explicitly religious point of view, is the Highlander Center. Highlander, which was a key actor in the civil rights movement, focuses on educating local leaders, helping them draw connections between the issues that affect them and better understand the power dynamics of those issues.[26]

Another example is the work of the late Helen Matthews Lewis, a scholar and activist who empowered residents of depressed Appalachian communities to apply critical political and theological reflection to their lived experiences.[27] In just one example of her extensive work, Lewis and scholar Mary Ann Hinsdale invited residents of Ivanhoe, Virginia, to utilize the insights of Latin American liberation theology to reflect on their town's economic decline and seek empowering and transformative responses.

Finally, Allen Johnson of CFTM draws on biblical conceptions of power to understand the power dynamics of MTR more clearly. He believes that viewing the operations of the coal industry in light of the biblical terminology of "powers and principalities" provides insight into how best to respond. For example, Walter Wink's theology of the powers argues that Jesus offers specific techniques for confronting the institutionalized system of domination in society. Wink claims that nonviolent acts such as turning the other cheek have a decidedly defiant character, and Johnson believes a similar sort of resourcefulness is called for in Appalachia.[28]

RESPONDING IN PARTICULAR SITUATIONS

A key aspect of theocentrism, one highlighted by both Niebuhr and Gustafson, is the belief that ethical decisions cannot be made in the abstract; they can be made only in particular concrete situations. According to Gustafson, "The Christian moral life is lived in response to God's actions in the situations in which the self exists. One confronts the creative, governing, and redeeming work of God in a concrete place."[29] Niebuhr reminds us that the unifying theocentric imagination is primarily a form of practical reasoning: it encounters God the creator and valuer not in the abstract but in concrete actions and relations.[30] This response in particular situations is a key moment in the pattern for theocentric moral action.

Integral to Niebuhr's view of Christian ethics as practical reasoning is an insistence on action. Responding to all actions as the actions of God means actually responding—that is, acting. As many liberation theologians have insisted, action is not the end result of a long process of discernment and reflection; rather, it is part of that discernment. The community comes to understand and clarify its principles only by acting on them. As Gustafson argues, principles become socially effective only when they are enacted and internalized by persons acting in decisive situations.[31] Thus, reflection and action are both parts of discernment. Niebuhr's emphasis on interpretation and imagination, an emphasis echoed in this book, does not mean that these reflective processes represent the only or the primary moment in his ethics. It is significant that Niebuhr offers responses in emergencies and responses to suffering as the paradigmatic examples of moral action as response.[32] It is precisely in these responses, where endless reflection is impossible and immediate action is called for, that the nature of morality is revealed most clearly. These decisions are shaped by the worldviews in which they are understood, but they also shape those worldviews. God's purposes are revealed not purely in reflection or interpretation but rather in the dialectic of interpretation and action.[33]

Practically speaking, this means that a key locus of ethics in the church is the congregation. This is where revelation and beliefs are brought to bear most directly on social action, as Gustafson points out.[34] However, I would argue that the congregation—typically understood as a stable local body of the institutional church—may not necessarily be the most appropriate place to think about this dialectic of interpretation and action. What is most important is that this process take place in a particular and appropriately small-scale community that is attempting to respond to a specific issue. Together, this community finds concrete ways to respond to a particular situation, and in those responses and the interpretation of them, the community progressively discerns and internalizes the interpretive framework in which it understands its actions. Here, the example of Roy Gene Crist, an Episcopal priest, and one of his congregations is informative.[35] Concerned about the immediate effects of a nearby mine, particularly the risks posed by overloaded and fast-moving coal trucks, some members of Crist's congregation sought to limit the practice of MTR. Like many other coalfield congregations, Crist's is far from unanimous in its feelings about MTR. But he and his parishioners have sought ways to engage one another as well as those outside the congregation; for example, the blessing services described earlier, which have led to some real dialogues, were Crist's idea. Crist and his congregation, like countless other mountain communities, are discerning a divine purpose not in the abstract but in response to particular situations that affect their lives.[36]

CULTIVATION

Addressing particular issues does not mean doing so in isolation or focusing an entire community's efforts on a single issue. Fundamental to theocentrism is the insight that values are inextricably interrelated and that ethical decisions cannot be considered in isolation from one another. Another facet of relationality, however, is equally essential to the moral life. In Niebuhr's understanding, the self (or the community of selves) finds coherence and continuity in the interpretive whole into which it fits its decisions.[37] The revelation that transforms this whole and all the values related to it thus transforms the entire life of the self or the community, and it does so not in a disinterested, objective way but as the revelation of a person who is both a source of value and an object of trust.[38] All values and all decisions are taken up into this transformation. At the same time, while Niebuhr speaks of this transformative revelation as if it were instantaneous, it is clearly a process of discernment: the divine purpose behind all actions to which one responds reveals itself only gradually, in a whole series of actions, responses, and interpretations.[39]

What this means for moral action in a community is that the selves making up that community reshape their being and their agency through the decisions, commitments, and communication of a common life organized around one shared center of value.[40] There is perhaps no more appropriate word for this process than the one Gustafson uses: *cultivation*. The community cultivates a moral life in its entire history of decisions and responses. In this process, it cultivates a partial understanding of the purposes of God to which it seeks to conform its actions. In other words, theocentric moral action requires ongoing efforts to respond theocentrically in the whole life of the community.

The Catholic Committee of Appalachia illustrates this practice of cultivation. The committee has been active in Appalachia for forty years and has addressed a wide range of issues, based partly on the theology of the two pastoral letters discussed in earlier chapters: *This Land Is Home to Me* and *At Home in the Web of Life*.[41] The ongoing process of reflection and action in the committee's work can be seen in the number of issues it has taken up and in the progressive discernment expressed in the letters and in a third document, *Models of Ministry*, which evaluates the success of the church's responses.[42] These three documents consciously build on one another and on conversations with Appalachians, gradually shaping a theology and ethics in response to this engagement. In this, they reveal a community engaged in a process of discernment through ongoing cultivation in the life of the church.

FAITH

Besides humility, the other principal attitude that characterizes Niebuhr's ethic is faith, defined as trust in the goodness of the principle of being.[43] This faith is the fundamental orientation that shapes and interprets all aspects of the moral life. Niebuhr argues that, in contrast to the distrust that finds meaning only in the self, "should it happen that confidence is given to me in the power by which all things are and by which I am; should I learn in the depths of my existence to praise the creative sources, then I shall understand and see that, *whatever is, is good,* affirmed by the power of being . . . and now all my relative evaluations will be subjected to the continuing and great correction."[44] Thus, in the church's effort to discern and respond to God's purposive action in creation, it turns finally to an attitude of confidence and trust that the principle of being and the source of all value is good and is active in creation. Like humility and repentance, this attitude is not abstract; it is made concrete in the prayers, reflections, and actions of the community.

Yet the God that is affirmed in this attitude is one whose values are not our own; the good that God seeks is not necessarily our own good.[45] Faith in this God is not the easy confidence that the good we recognize will ultimately triumph; rather, it is the belief that, because God is benevolent, good will triumph, whether we recognize it as good or not. It is the prayer, "Thy will, not mine, be done."[46] There are trust and security, to be sure, in the affirmation of a unified and benevolent will acting in all creation. There is, however, little reassurance.

I have proposed these six practices as a pattern for theocentric moral action. These practices can, I believe, enact the process of discernment in which

God's purposive action in creation is revealed and responded to. It is an ongoing dialectic of response and interpretation. The church never has a full grasp of God's purposes; it can only glimpse them incompletely. Appropriately, Niebuhr compares this progressive revelation to a mountain:

> We climb the mountain of revelation that we may gain a view of the shadowed valley in which we dwell and from the valley we look up again to the mountain. Each arduous journey brings new understanding, but also new wonder and surprise. This mountain is not one we climbed once upon a time; it is a well-known peak we never wholly know, which must be climbed again in every generation, on every new day. There is no time or place in human history, there is no moment in the church's past, nor is there any set of doctrines, any philosophy or theology of which we might say, "Here the knowledge possible through revelation and the knowledge of revelation is fully set forth." Revelation is not only progressive but it requires of those to whom it has come that they begin the never-ending pilgrim's progress of the reasoning Christian heart.[47]

A Relationship to Place

We return now to the relationship to place, not as a general characteristic of Appalachia and Appalachians, as in chapter 4, but as a feature of the theocentric approach. Given Niebuhr's emphasis on the contextual, concrete, relational character of moral action, this approach cannot ignore the fact that moral action always occurs in a place. Certainly, discussions of MTR remind us that moral issues are inextricable from the places where they occur: as noted in chapter 4, questions of power, identity, and destruction all incorporate ideas about people's relationship to place. In this section, I argue that the theocentric approach needs to be complemented by careful attention to place as a locus of God's action in the world; this idea can be described with the help of the popular concept of "loving the mountains." I begin, however, by returning to the criticism that the theocentric approach is too quiescent in the face of injustice. I contend that, when properly complemented by this relationship to place, this approach can indeed support a strong challenge to unjust or exploitative practices. I then conclude with what may be the most promising ground for such a challenge—the notion of an Anthropocene epoch.

ABSOLUTIZING THE RELATIVE

Recall that for several reasons—its strong emphasis on divine agency and its downplaying of human agency, the related attitudes of humility and trust in the benevolent principle of being, the belief that God is acting in all creaturely actions, and the correlative exhortation to consider innumerable details and varying perspectives—the theocentric view might be seen as passive or complacent. Again, Niebuhr himself was aware of this challenge: "Does not this way of understanding . . . the model Christian life . . . make of this life an affair of pure resignation to the will of God?"[48] Seen one way, this criticism represents the absolutization of the relative against which Niebuhr argues so forcefully.

At the beginning of *The Meaning of Revelation,* Niebuhr argues that the great source evil in life is "absolutizing the relative," which in Christianity means substituting religion, the church, or morality for God.[49] He explains that theology may take the standpoint of either faith in God or faith in something else, be it religion, morality, humanity, or civilization.[50] Any time theology limits God's self-revelation based on human values or interests, which necessarily includes the values of any ethical standpoint, it ceases to be theocentric and becomes anthropocentric and is guilty of absolutizing what ought to be relative.

Thus, the criticism that Niebuhr's approach may lead to passivity in the face of injustice is to some extent correct, but it misses the force of his argument. The criticism itself presupposes some notion of justice and imposes this as a condition for a legitimate ethic: if a theocentric approach (or any other approach) cannot adequately support a prophetic denunciation of injustice so conceived, this is viewed as a weakness. This is precisely what Niebuhr warns against: imposing human values on divine revelation. If careful attention to the multifaceted revelation of God's will in a complicated situation is preempted by a perceived need to decry certain unjust situations, this is an anthropocentric imagination. "The God who is primarily a helper toward the attainment of human wishes," he argues, "is not the being to whom Christ said, 'Thy will, not mine, be done.'"[51] Surely this includes the wish to see justice (as we perceive it) done and injustice stopped. If, in the end, this means that a consistent application of Niebuhr's approach is less capable of challenging perceived injustices because they are viewed as more ambiguous than previously imagined, this must be acknowledged as part of the difficulty of radical monotheism.[52] Conversely, advocates of a preconceived idea of justice may argue that the sacrifice of a forceful ethical critique of injustice is too high a price to pay for monotheistic consistency. Although this is a compelling claim, a theocentric perspective would obviously disagree.

Yet there is another sense in which this challenge misses the force of Niebuhr's position. When Niebuhr argues that Christian ethics ought to be characterized by submission, rather than freedom, he is not advocating a straightforward ethic of submission to God's will, as exemplified in some divine command theories of ethics.[53] As discussed in chapter 3, Niebuhr has a place for free human agency—namely, in the choosing and modifying of interpretations. Thus, the submission he speaks of is not complete abdication of the responsibility to act; rather, it is the conforming of our interpretations and actions to the purposes of God, to the extent possible. As Gustafson rightly insists, our ability to act means that we are accountable for our actions (or inaction), even though we can never fully know the telos toward which our actions are directed.

Thus, far from undermining a strong prophetic stance, the theocentric approach offers a stronger foundation for such a stance. Ultimately, beneath the circumspection and careful discernment, this approach's central claims are that God has a purpose, that this purpose is revealed to persons in some limited way, and that believers must respond to and affirm this purpose in their actions. As Niebuhr argues convincingly, this self-consciously objective foundation for human moral action is more secure than the shifting and multiple value centers of polytheistic and anthropocentric perspectives. According to Niebuhr's anthropology, consistency and objectivity are found only in monotheism; other systems of value, including those based on ideas of justice, are ultimately unsustainable because they are built around relative centers mistakenly viewed as absolute. Seen this way, rather than weakening a prophetic call for justice, the theocentric approach can actually strengthen it. At the same time, discerning the meaning of justice in the context of divine, rather than human, purposes requires finding and cultivating the right imaginations.

THE VOICE OF THE MOUNTAINS

Niebuhr's main response to the criticism that his approach leads to resignation is not the one I just described—that such a challenge misrepresents his position. Rather, he argues that fatalism uses the wrong images to interpret situations. God is not the pure determiner of destiny but the loving father and dynamic ruler. "Since we shall in any case use myths," he affirms, "let us use our myths critically and with discrimination."[54] On the surface, this seems to beg the question: presumably, the determination of which images or myths are most suitable is exactly the issue at hand.

Elsewhere, however, Niebuhr argues that destructive or inadequate imaginations are known by their fruits: alienation, destruction, conflict.[55] To avoid being simply tautological, this argument must indicate an experiential reference for moral discernment. That is, even though finite human values cannot be permitted to stand in for the absolute value, we can recognize values and, perhaps more clearly, disvalues in our experience. Only the light of revelation allows us to understand these disvalues for what they are—the product of self-centered imaginations. Yet we can intuitively grasp their destructiveness.[56]

Gustafson illuminates the role of experience more thoroughly. Similar to his approach in *Ethics from a Theocentric Perspective,* his theocentric approach to the environment in *A Sense of the Divine* focuses on understanding and deepening that sense, which he believes is part of the human experience of nature.[57] Although there is no pure experience unmediated by cultural values and narratives, humans' stances toward nature are always based on their experiences of it. We experience nature on a variety of levels, from the most intimate, tactile encounter to an awestruck sense of our place on the planet. And although the precise identification of what constitutes value or disvalue, beauty or ugliness, is always difficult, he argues that we perceive the threat of evil more readily than we grasp what is good, that we recognize disvalue more clearly than we understand value. We can perceive threats to the well-being of our species or creation more readily than we can articulate what well-being requires.[58] Human revulsion at certain potential disvalues, he believes, can therefore be a key indicator of what we value and how we value it. Our immediate perception of ugliness in a landscape, for example, can help us articulate what beauty might entail. Combined with Niebuhr's ideas about imaginations and their fruits, this suggests that the experience of environmental disvalue may be, like the experience of other kinds of alienation, an indication of destructive imaginations.[59]

This experiential reference for the theocentric approach leaves some room for a strong prophetic stance. In spite of this approach's insistence on the relativism of all finite values, it retains the ability to challenge certain practices within a theocentric framework based on the experience of disvalue. After all, central to theocentrism is the belief that God's purposes are expressed in all parts of creation. Niebuhr argues that, in this respect, his monotheism is more comprehensive than humanism, Albert Schweitzer's reverence for life, or even naturalism.[60] Although this book has focused on the human imaginations that may conceal or reveal traces of divinity, certainly there must be expressions of God's action that can be experienced in creation itself, in what Gustafson calls a "pre-reflexive" way.[61] Put differently, among the multiple voices to which an agent responds, surely there must be those of particular places themselves: the voices of nonhuman creatures, soil, water, or mountains.[62]

There is a popular bumper sticker distributed by the Highlands Conservancy in West Virginia that reads, "I [Heart] Mountains." Properly understood, this sentiment can be a useful way of thinking about how we might listen to the mountains' voices as an expression of God's will in creation. If discerning justice from a theocentric perspective requires selecting and applying the right kinds of imaginations, I submit that loving the mountains represents an appropriate way of imagining the right human relationships with the places of Appalachia, one that incorporates and elaborates Gustafson's sense of the divine and the intuitive perception of value and disvalue, ambiguity and dependence, in nature. Obviously, this kind of prereflexive or intuitive experience of nature resists clear description, and in any case, a comprehensive treatment of this topic is beyond the scope of this study. Nonetheless, I contend that loving the mountains—understood as a careful and active attention to the particular places of Appalachia—is an apt model for discerning God's will as expressed in creation itself.

Philosopher Iris Murdoch describes the moral life as a carefully cultivated attention to particular individuals, "a patient and just discernment and exploration of what confronts one, which is the result not simply of opening one's eyes but of a certainly perfectly familiar kind of moral discipline."[63] Goodness, she says, is known not in the abstract but only in the patient attention to individuals that she identifies as love.[64] In terms quite similar to the dichotomy of self-centered versus God-centered imaginations, Murdoch argues that love is the practice of "really looking," attending to a particular reality without "returning surreptitiously to the self with consolations of self-pity, resentment, fantasy and despair."[65] This kind of love, she argues, is inseparable from justice. She offers as a paradigmatic example attention to nature, "a self-forgetful pleasure in the sheer alien pointless independent existence of animals, birds, stones and trees."[66]

Murdoch's description of "the sheer alien pointless independent existence" of nature risks imagining it as pristine, or imagining our experience of it as unmediated by our own beliefs and images. But as I have noted repeatedly, there is no such experience; as Gustafson observes, every encounter with nature already involves the "cultural penetrations of nature."[67] Further, human beings inevitably interact with their environment. The pristine wilderness witnessed by an uninvolved observer is a fantasy. If, however, we understand Murdoch's love to include a loving interaction with nature— for example, the careful planting of wildflowers at a mountain-blessing service—this can be a model for the kind of attention needed to hear the voices of the mountains. Murdoch's invocation of the Platonic notion of craft (techné) as a way of cultivating careful attention suggests this interpretation.[68] By learning and practicing a craft, one moves away from self-concern toward a clear-eyed appreciation of the reality of the thing that is loved. Examples of crafts that inform and strengthen a love of Appalachia might include farming, harvesting ginseng in the mountains, and studying firsthand the biology of the region. Even deep mining might represent an attentive connection to the mountains that carries one away from selfish interests and allows a place to speak. It is hard to imagine a more intimate experience of the mountains than that of underground mining, and to deny that this could be a craft through which one becomes more attuned to the voice of the mountains seems prejudiced.[69] In any case, the point is that we can discern God's action in a place only by loving it in Murdoch's sense, and we can love a place only through attentive and careful interaction with it.

There may be no better exponent of this kind of active, loving attention to a place than Wendell Berry.[70] Berry poignantly describes the patience and discipline required to stand in a place and ask of it, "What is this place? What is in it? What is its nature? How should men live in it? What must I do?"[71] The answers come only gradually and partially, if at all; it is the asking itself, the humility required to learn from a place, that is essential. This discipline leads to the perseverance and skill needed to use and care for this part of creation.[72] Berry's ethic is not a spectator's appreciation of pristine landscape. His description of the proper relationship to a place is an eloquent expression of loving attention, and it is worth quoting at length:

> Only in . . . silence and darkness can [a man] recover the sense of the world's longevity, of its ability to thrive without him, of his inferiority to it and his dependence on it. Perhaps then, having heard that silence and seen that darkness, he will grow humble before

the place and begin to take it in—to learn *from it* what it is. As its sounds come into his hearing, and its lights and colors come into his vision, and its odors come into his nostrils, then he may come into *its* presence as he never has before, and he will arrive in his place and will want to remain.[73]

This is a demanding vision, yet surely this is what it means to truly love the mountains.

This image of loving the mountains does not reinscribe the insider-outsider duality I criticized in earlier chapters. I am arguing that actively knowing and interacting with a place is essential to discerning God's action and purposes in that place. The focus is on practices of attention and care. This is not an innate connection to place; rather, it is a practiced way of being in a place. It recognizes and is critical of the narratives and interpretations that implicitly separate people from one another as well as from their places.

What I am suggesting, then, is that loving the mountains in this way is a necessary complement to the theocentric approach described in this book, a necessary part of an overall ethic of theocentric action. As Gustafson argues, there is no experience of the natural environment that is unmediated by cultural ideas and images, yet the experience of nature in a place is a fundamental touchstone for our moral positions regarding the world. In these encounters, we have intuitions of value and especially of disvalue, and although these intuitions can be scrutinized and evaluated in light of our central values, they remain important to our attempt to discern the will of the creator. They can inform which imaginations we choose and which we reject. We can listen for the voice of the mountains among the multitude of voices to which we respond. What is more, we can develop our ability to listen through active, caring, attentive interaction. Of course, this is not done in the abstract. The sounds, lights, colors, and odors of a place cannot be known in a general way. We cannot love the mountains without actually being in the mountains from time to time.

THEOCENTRISM IN THE ANTHROPOCENE EPOCH

Some environmental scientists speak of the current age as a new "Anthropocene epoch."[74] These scholars argue that the radical and pervasive impacts of human action on the natural world are so significant—dramatically transforming nearly all aspects of the land, water, and air around us—that they have given rise to a new geological age subsequent to the Holocene epoch and that began at the end of the eighteenth century. In proposing the term Anthropocene epoch, Nobel laureate and atmospheric chemist Paul Crutzen cites a litany of such impacts: exhausting in a few generations fossil fuels that took several hundred million years to create, transforming 30 to 50 percent of the earth's terrain, using more than half of all accessible freshwater, and increasing the species extinction rate by as much as 10,000 times.[75] As Willis Jenkins points out, this notion of an epoch defined by human action creates some novel challenges for ethics. It forces us to recognize our responsibility—in both a causal and a moral sense—for these ambiguous impacts and come up with a plan to manage them. Human manipulation of the environment is not, he notes, wrong in itself, but it does impose a new degree of accountability on human agents.[76]

Perhaps no single practice is as emblematic of this new epoch as MTR. The sheer force and scale of MTR's effect on the earth are unique. The infamous dragline, for example, which allows a single operator to move 100,000 tons of earth in one scoop, is a powerful symbol of humanity's unprecedented ability to shape its environment. Note, too, that MTR is associated with all the dramatic human impacts cited in the previous paragraph. Relative to the aggregate changes our species has effected on the earth, MTR is only one minor factor. Yet it is difficult to imagine another process that so clearly represents the massive transformations of the Anthropocene epoch.

From a theocentric perspective, the Anthropocene epoch epitomizes the pathological propensity to place human interests at the center of our value systems. Of course, human activities that reorganize entire ecosystems do not irrefutably prove the presence of human-centered imaginations. Taken as a whole, however, especially given the scope and pervasiveness of human impacts, the Anthropocene clearly suggests an entire age of humanity that has failed to properly orient its values around God (or any other nonhuman value). Reinhold Niebuhr famously asserted that the only empirically verifiable Christian doctrine is original sin; the Anthropocene epoch seems to come as close as possible to empirically verifying his brother's claims about the human inclination toward idolatry and absolutizing what ought to be relative.

From the theocentric perspective, then, MTR is uniquely indicative of an arrogation of God's proper place at the center of human imaginations, and as such, it ought to be stopped. If this clear denunciation seems inconsistent with my heretofore moderate, nuanced approach, I would argue that it is not. Condemning MTR as representative of an idolatrous imagination is different from claiming that the practice is wrong in itself. The key difference is that the latter claim would lay blame at the feet of the owners, operators, and miners who practice MTR, while the former claim views MTR as part of a larger pathology of human-centered imaginations. MTR is not itself idolatrous; rather, it is a symptom of an idolatry that is both deeper and broader, one that has come to define its own geological era.

It is possible to draw a comparison with Niebuhr's discussion of war as God's judgment. Like war, MTR indisputably causes some disvalue to biological communities (although, as I have shown, the amount of harm is still unclear). Like war, this disvalue falls disproportionately on those communities with little or no responsibility for the practice, whether human or nonhuman. Therefore, like war, MTR stands as a judgment not on those who are harmed by it, nor even on those who are considered aggressors in it; rather, it is a judgment on the self-centered society that allows—even requires—such a practice.[77] Responding to this interpretation of MTR calls for abandoning attitudes of self-defense and indignation and working to protect the victims—the same response that Niebuhr urges with respect to war.

Thus, even as a theocentric approach condemns MTR as a symbol of an idolatrous age, the appropriate response is more complicated than simply ending the practice. Vilifying MTR and those involved in it bypasses the nuanced questions of identity and the social construction of values and does little to challenge human-centered imaginations. Indeed, in some instances, it seems to be a matter of replacing one set of human-centered values with another. Instead, a theocentric approach to MTR—and the imaginations that support it—suggests a somewhat more modest ethical response.

Moving Forward: Tentative Guidelines

Up to this point, my approach has focused on attitudes and practices rather than specific actions; that is, it has been more "agent ethics" than "act ethics." I have avoided making general statements about the morality of MTR because I believe, like Niebuhr, that, when abstracted from the context of concrete relationships of value, such generalizations are unhelpful, misguided, or both. As I mentioned in chapter 3, I am less optimistic than Gustafson about the general legibility of the order of the world: it is clear, from both Niebuhr's work on responsibility and relationality and the sociological work I have drawn on throughout this book, that the relationships that give order to social life and to issues like MTR can be interpreted in myriad ways. For this reason, guidelines about the morality of certain actions must be discerned in particular situations, oriented by the kinds of practices and narratives I have described.

Nonetheless, with this contextual foundation firmly in place, and with due recognition of the ambiguity of most moral decision making, I believe that the approach I have described should provide some general guidelines. Again, these are not clear, specific mandates that can be read inductively off of the created order. Rather, they are general guidelines that emerge from the dialectic of revelation that is fundamental to theocentrism: revelation—specifically, the event of Jesus Christ—illuminates experience, which in turn clarifies revelation. I offer four general guidelines and suggest some implications of each.

First, the mountains must be protected and cared for, since they are interpreted both as valued by God and as in some way revelatory of God's purposes. If this does not require an outright and immediate prohibition of MTR, for the reasons noted above, it at least entails a strong presumption against it and a prohibition of the reckless ways in which it is often practiced.[78] Policies established to protect different aspects of the mountain ecosystem need to be enforced, not circumvented, by agencies that are more accountable to persons and communities than to corporations and politicians, and if such policies are ineffective or unclear, they need to be modified. For example, the introduction mentioned reclamation bonds, which are forfeited if reclamation does not meet government standards and used to fund adequate reclamation. This requirement should be strictly enforced and, if necessary, strengthened.[79] Mine companies and enforcement agencies alike should institute reclamation practices that restore the premining appearance and function of mined areas, to the extent possible, in both new and previously mined, unreclaimed sites. Approximate original contour (AOC) variances need to be the subject of real deliberation rather than a matter of expediency for the industry. And parties on both sides need to do more to explore, develop, and demand careful reclamation.

Second, and a necessary corollary to caring for the mountains, is greater attention to the details, in particular to scientific data, and a call for more complete research. In chapter 1, I shifted the focus of this study from those scientific data to the imaginations that surround and shape them. Yet a key characteristic of theocentrism—seen equally in Niebuhr's "What is going on?" and in Gustafson's "God is in the details"—is the belief that because God acts through all creation, the sciences are potentially revelatory. According to Niebuhr, "a genuinely disinterested science may be one of the greatest affirmations of faith."[80] Thus, the theocentric perspective embraces scientific insights and laments the incomplete data surrounding MTR. Research that proves (or disproves) causal links, rather than simple correlation, between MTR and some of its adverse effects is urgently needed. Likewise, clearer data about the economic impacts of MTR are desirable. In particular, the forestry reclamation approach described in chapter 1 should be cautiously encouraged, and the impressive work of those scholars and institutions mentioned in chapter 5 (Samir Doshi and the University of Kentucky, for example) should be followed carefully. Finally, a theocentric perspective supports the requirement of individual permits and individual environmental impact statements for each new mine (see chapter 1), regulations that have been affirmed by district courts only to be repeatedly overturned by the Fourth Circuit Court of Appeals. These requirements would allow a more complete understanding of how God may be acting in this situation. Again, this careful attention to scientific data issues from the theocentric approach not as prior to or independent of an examination of imaginations but as integral to it—complementary efforts to discern what is going on.

The mountains and their ecosystems are not the only values invoked in this debate, and even the value of the mountains can be multivalent. Thus, the third guideline involves the avoidance of polarizing rhetoric and dramatic protests, which have too often characterized this debate. The inadequacy of these forms of engagement, both theologically and morally, has been a major theme of this study and does not require further elaboration.

Fourth, given the multiplicity and relationality of the values involved in MTR, a more expansive conception of the relevant wholes is required. In Gustafson's terms, this means a broader perspective that encompasses the entire regional economy rather than a single industry, one that looks beyond the needs of the current generation to include future generations. This entails more comprehensive efforts to find both economic and energy alternatives, placing the debate about MTR in its broader contexts as described in chapter 5. People on both sides of this debate recognize that efforts to minimize the effects and extent of MTR and surface mining need to be accompanied by the development of new energy sources for the nation and new revenue sources for a region that has so far resisted such diversification. Of course, all the various possibilities involve their own competing values and narratives and therefore require equally careful processes of engagement and discernment.

The relationships in which values are negotiated are multiple and dynamic, and they are continually relativized in light of the progressive discernment of divine purposes. As Gustafson affirms, a theocentric approach acknowledges that it cannot offer a fixed hierarchy of values or inviolable principles for moral decision making.[81] To try to do so would be to succumb to the perennial temptation to substitute human beings and their interests for God. Thus, how to apply even such general guidelines as those I have offered is a matter for particular communities of persons in particular situations to decide.

Ultimately, however, the theocentric approach I have described, grounded by a practiced love for the mountains, can support a strong critique of MTR and a judgment of the Anthropocene epoch. It can say, along with AMERC director Lon Oliver, that some things need to be stopped, but that most people go there too quickly.[82] We, as a society, can do better than MTR; certain aspects of the practice itself, the debate around it, and the imaginations that support it are contrary to God's purposes in creation, sometimes starkly so. Strong prophetic voices are needed to recall the community to a vision of those purposes. Yet unless those voices and that vision come from a knowledge of the places themselves, from what I call a true love of the mountains, they will ring false.

Many of those already mentioned in this study exemplify some aspects of the overall approach I have been describing: Lon Oliver and AMERC illustrate the humble discernment of God's action in the world. Allen Johnson and Christians for the Mountains exemplify the Niebuhrian attitudes of confidence and hope rooted in the belief that God is acting in the world. Mine owner Andrew Jordon represents openness to real conversation. And perhaps no one better exemplifies the deep love of and careful attention to the mountains I have just described than the late Larry Gibson. Gibson dedicated his life to protecting his family home on Kayford Mountain in southern West Virginia. He traveled around the country, speaking about and raising awareness of MTR, and he brought hundreds of visitors to his home to see MTR's effects firsthand.

Yet the best illustration of the theocentric approach may be the less remarkable, gentler ministries of Roy Gene Crist and Stan Holmes, two Episcopal priests serving congregations in southern West Virginia. Holmes and Crist were instrumental in conceiving and planning the mountain-blessing services described earlier, which they see as a way to instill greater reverence for the blessedness of the entire region. They have also educated others, both in their congregations and on the larger diocesan level, about the facts of MTR, seeking a more respectful dialogue that eschews ideologies and searches for practical solutions. As parish priests, they minister to people on both sides of the issue, and they are judicious about the stances they take. As Holmes argues, it would be unconscionable to exclude supporters of MTR from the church community because of his own opposition to the practice. A former coal company supervisor, he knows the passions on both sides of the debate, as well as the fear and pride that motivate defenders of MTR. At the same time, his congregants are well aware of his position. He has preached about the challenges facing the region, although he does not preach explicitly anti-MTR sermons. Some church members have been directly involved in political movements to mitigate MTR's effects on their communities, and they have been encouraged by their success. Yet the priests themselves avoid activism, for both practical and theological reasons. Holmes hopes to encourage real conversation about how to mine responsibly.[83] In their attitudes of humility and hope, their openness to real dialogue, their focus on practical concerns, and their efforts to nurture and cultivate their congregations' engagement, these leaders and their communities illustrate the overall approach I have described. Certainly, theirs is an imperfect example. This incompleteness, however, is in keeping with Niebuhr's belief that the monotheism he advocates emerges "more as hope than as datum, more perhaps as a possibility than as an actuality"; yet it occasionally emerges nonetheless.[84]

This is the kind of response a theocentric approach to MTR supports. It is characterized by humility and hope, rather than certainty. It tends toward details and gradations, rather than bold denunciations. It scrutinizes the narratives and images used to interpret the issue and finds and cultivates theocentric imaginations to replace self-centered ones. It seeks the lost, those voices left out of dominant discourses, with the conviction that God's purposes are expressed in all the actions and interpretations of creation.[85] It speaks from a real, active, disciplined love of the mountains. And it is precisely because of this careful discernment and reflection, not in spite of it, that this approach can confront destructive and alienating imaginations on all sides, not superficially but radically, assailing their very foundations. It can, with humility and confidence, challenge believers to turn their eyes to the actions and purposes of God. It can proclaim in the words of the psalmist, the words hand-painted on a stone at the entrance to a mountaintop family cemetary in West Virginia: "In his hand are the depths of the earth; the heights of the mountains are his also."

Acknowledgments

I am grateful to all those who helped this project come to fruition. They include the professors and colleagues who read and commented on drafts of the manuscript: Margaret Farley, Willis Jenkins, Thomas Ogletree, Frederick Simmons, Emilie Townes, Dwight Billings, Stan Brunn, Timothy Hiller, and the anonymous readers for the University Press of Kentucky; all those in West Virginia, Kentucky, and Tennessee who shared their work and stories with me: Corinne Almquist, Roy Gene Crist, Larry Gibson, Stan Holmes, Pat Hudson, Allen Johnson, Andrew Jordon, Lon Oliver, John Rausch, Jim Sessions, Sage Vekasi-Phillips, and Carol Warren; Vivian Stockman, for sharing her striking photos of MTR; and the University Press of Kentucky. Finally, I could have done nothing without the support and encouragement of my wife, Leigh Preston, or the love of my two sons, Cabell and Cullen; I am especially grateful to them, and for them.

Notes

Preface

1. Edward S. Casey, *Getting Back into Place: Toward a Renewed Understanding of the Place-World* (Bloomington: Indiana University Press, 2009), 319–23.

Introduction

1. S. P. Hanna, "Representation and the Reproduction of Appalachian Space: A History of Contested Signs and Meanings," *Historical Geography* 28 (2000): 179–207.

2. Thomas James, "Responsibility Ethics and Postliberalism: Rereading H. Richard Niebuhr's *The Meaning of Revelation*," *Political Theology* 13, no. 1 (December 4, 2012): 44.

3. Willis Jenkins, *The Future of Ethics: Sustainability, Social Justice, and Religious Creativity* (Washington, DC: Georgetown University Press, 2013), 105.

4. H. Richard Niebuhr, *The Responsible Self: An Essay in Christian Moral Philosophy* (New York: Harper & Row, 1963), 45–46.

5. In this section and the one that follows, I draw on sources from a variety of fields, including social scientists (Shirley Stewart Burns, Rebecca Scott), journalists (Michael Shnayerson, Erik Reece), natural scientists (Nathaniel Hitt, Michael Hendryx, M. A. Palmer), and activists (Silas House, Jack Spadaro).

6. West Virginia Coal Association, *West Virginia Coal: Fueling an American Renaissance (Coal Facts 2011)* (Charleston: West Virginia Coal Association, 2011), 46.

7. Shirley Stewart Burns, *Bringing Down the Mountains: The Impact of Mountaintop Removal on Southern West Virginia Communities* (Morgantown: West Virginia University Press, 2007), 5–6.

8. Ibid.; Rebecca R. Scott, *Removing Mountains: Extracting Nature and Identity in the Appalachian Coalfields* (Minneapolis: University of Minnesota Press, 2010), 81; Michael Shnayerson, *Coal River* (New York: Farrar, Straus & Giroux, 2008), 63.

9. Scott, *Removing Mountains,* 81.

10. West Virginia Coal Association, *West Virginia Coal,* 33.

11. Shnayerson, *Coal River,* 102.

12. Rory McIlmoil, Evan Hansen, Ted Boettner, and Paul Miller, "Coal and Renewables in Central Appalachia: The Impact of Coal on the West Virginia State Budget" (Downstream Strategies and WV Center on Budget and Policy, June 22, 2010).

13. Andrew Jordon, interview with the author, Charleston, WV, August 9, 2010; West Virginia Coal Association, *West Virginia Coal,* 36; *Reclaiming the Future: Reforestation in Appalachia* (Lexington: University of Kentucky College of Agriculture, Kentucky State University, 2008), film.

14. Jordon interview.

15. Silas House, *Something's Rising: Appalachians Fighting Mountaintop Removal* (Lexington: University Press of Kentucky, 2009), 2.

16. Burns, *Bringing Down the Mountains*, 6.

17. Ibid., 14.

18. Appalachian Voices, "Update: Extent of Mountaintop Mining in Appalachia as of 2008," http://ilovemountains.org/reclamation-fail/details.php#extent_study_2012.

19. Jack Spadaro, "Mountaintop Removal: The Destruction of Appalachia," in *Plundering Appalachia: The Tragedy of Mountaintop-Removal Coal Mining*, ed. Tom Butler and George Wuerthner (San Rafael, CA: Earth Aware, 2009), 59.

20. M. A. Palmer et al., "Mountaintop Mining Consequences," *Science* 327, no. 5962 (January 8, 2010): 148–49, doi:10.1126/science.1180543; Butler and Wuerthner, *Plundering Appalachia*, 3, 60.

21. George Wuerthner, "Appalachia: Land of Diversity," in *Plundering Appalachia*, 3; Erik Reece, *Lost Mountain: A Year in the Vanishing Wilderness: Radical Strip Mining and the Devastation of Appalachia* (New York: Riverhead Books, 2006), 35–36.

22. Wuerthner, "Appalachia: Land of Diversity," 3.

23. Butler and Wuerthner, *Plundering Appalachia*, 50.

24. Palmer et al., "Mountaintop Mining Consequences."

25. Ibid.

26. Ibid.

27. Nathaniel P. Hitt and Michael Hendryx, "Ecological Integrity of Streams Related to Human Cancer Mortality Rates," *EcoHealth* 7, no. 1 (April 2010): 91–104.

28. Michael Hendryx and Keith J. Zullig, "Higher Coronary Heart Disease and Heart Attack Morbidity in Appalachian Coal Mining Regions," *Preventive Medicine* 49, no. 5 (November 2009): 355–59, doi:10.1016/j.ypmed.2009.09.011.

29. Palmer et al., "Mountaintop Mining Consequences."

30. Reece, *Lost Mountain*, 115–17.

31. Burns, *Bringing Down the Mountains*, 40–41, 141.

32. House, *Something's Rising*, 9.

33. Reece, *Lost Mountain*, 37–38.

34. West Virginia Coal Association, *West Virginia Coal*, 33.

35. Some explanation is therefore required for my own choice of terminology. I opt for "mountaintop removal" because most of the people and communities I discuss use this phrase. If this indicates a particular bias, that may be inevitable. Nonetheless, the ideas and assumptions implicit in such descriptors are the subject of critical examination in subsequent chapters.

36. House, *Something's Rising*.

37. Ibid.; Burns, *Bringing Down the Mountains*, 49–51.

38. House, *Something's Rising*, 137, 142.

39. Butler and Wuerthner, *Plundering Appalachia*, 85.

40. "Goldman Prize," http://www.goldmanprize.org/theprize/about (accessed March 5, 2013).

41. Butler and Wuerthner, *Plundering Appalachia,* 83.

42. Scott, *Removing Mountains,* 102–4.

43. Phylis Geller, *Coal Country* (Evening Star, 2009), film.

44. Larry Gibson, interview with the author, Kayford Mountain, WV, August 8, 2010.

45. Burns, *Bringing Down the Mountains,* 46–47.

46. Jeff Biggers, "Updated: VIDEO: Nonviolent Goldman Prize Winner Attacked by Massey Supporter: 94-Year-Old Hechler, Hannah, Hansen Arrested at Coal River," *Huffington Post,* http://www.huffingtonpost.com/jeff-biggers/live-at-coal-river-daryl_b_219628.html (accessed November 8, 2012).

47. It is worth noting that Jordon believes Blankenship has been unjustly vilified and cares as much about the safety of his workers and their communities as any other person would. Jordon interview.

48. Reece, *Lost Mountain;* Jeff Goodell, *Big Coal: The Dirty Secret behind America's Energy Future* (Boston: Houghton Mifflin, 2006).

49. Tom Zeller Jr., "A Battle in West Virginia Mining Country Pits Coal against Wind," *New York Times,* August 14, 2010, Business/Energy & Environment sec.; Amanda Paulson, "In Coal Country, Heat Rises over Latest Method of Mining," *Christian Science Monitor,* January 3, 2006; Erik Eckholm, "West Virginia Sues over Mountaintop Mining Limits," NYTimes.com, October 13, 2010.

50. Shnayerson, *Coal River;* House, *Something's Rising;* Penny Loeb, *Moving Mountains: How One Woman and Her Community Won Justice from Big Coal* (Lexington: University Press of Kentucky, 2007).

51. Bill Moyers, "Moyers on America. Is God Green?" PBS, 2006, http://www.pbs.org/moyers/moyersonamerica/green/index.html; Peter Smith, "Religion Shaping Mountain-top Removal Debate in Appalachia Coal Country," *Louisville Courier-Journal,* December 19, 2009; Associated Press, "W.Va. Churches Slam Proposed Mining Rule," *New York Times,* October 11, 2007; House, *Something's Rising;* Reece, *Lost Mountain.*

52. Palmer et al., "Mountaintop Mining Consequences"; Hendryx and Zullig, "Higher Coronary Heart Disease"; Hitt and Hendryx, "Ecological Integrity of Streams"; Charles E. Davis and Robert J. Duffy, "King Coal vs. Reclamation: Federal Regulation of Mountaintop Removal Mining in Appalachia," *Administration & Society* 41, no. 6 (October 1, 2009): 674–92.

53. Compare Benita Howell, *Culture, Environment, and Conservation in the Appalachian South* (Urbana: University of Illinois Press, 2002).

54. Harry M. Caudill, *Night Comes to the Cumberlands: A Biography of a Depressed Area* (Boston: Little, Brown, 1963). Compare Allen Batteau, *The Invention of Appalachia: The Anthropology of Form and Meaning* (Tucson: University of Arizona Press, 1990); Henry D. Shapiro, *Appalachia on Our Mind: The Southern Mountains and Mountaineers in the American Consciousness, 1870–1920* (Chapel Hill: University of North Carolina Press, 1978).

55. Batteau, *Invention of Appalachia.*

56. Scott, *Removing Mountains;* Michele Morrone and Geoffrey L. Buckley, eds., *Mountains of Injustice: Social and Environmental Justice in Appalachia* (Athens: Ohio University Press, 2011); Bryan T. McNeil, *Combating Mountaintop Removal: New Directions in the Fight against Big Coal* (Urbana: University of Illinois Press, 2011); Burns, *Bringing Down the Mountains.*

57. Batteau, *Invention of Appalachia.*

58. David E. Whisnant, *Modernizing the Mountaineer: People, Power, and Planning in Appalachia* (Boone, NC: Appalachian Consortium Press, 1980); David E. Whisnant, *All That Is Native & Fine: The Politics of Culture in an American Region* (Chapel Hill: University of North Carolina Press, 1983).

59. John Gaventa, *Power and Powerlessness: Quiescence and Rebellion in an Appalachian Valley* (Urbana: University of Illinois Press, 1980).

60. Stephen L. Fisher, ed., *Fighting Back in Appalachia: Traditions of Resistance and Change* (Philadelphia: Temple University Press, 1993); Chad Montrie, *To Save the Land and People: A History of Opposition to Surface Coal Mining in Appalachia* (Chapel Hill: University of North Carolina Press, 2003).

61. Batteau, *Invention of Appalachia;* Howell, *Culture, Environment, and Conservation.*

62. Dwight B. Billings and Shaunna L. Scott, "Religion and Political Legitimation," *Annual Review of Sociology* 20, no. 1 (1994): 173–201; Dwight B. Billings, "Religion as Opposition: A Gramscian Analysis," *American Journal of Sociology* 96, no. 1 (July 1990): 1–31; Dwight B. Billings and Will Samson, "Evangelical Christians and the Environment: 'Christians for the Mountains' and the Appalachian Movement against Mountaintop Removal Coal Mining," *Worldviews: Global Religions, Culture & Ecology* 16, no. 1 (2012): 1–29.

63. Richard Callahan, *Work and Faith in the Kentucky Coal Fields: Subject to Dust* (Bloomington: Indiana University Press, 2009); Fisher, *Fighting Back in Appalachia;* Bill Leonard, ed., *Christianity in Appalachia: Profiles in Regional Pluralism* (Knoxville: University of Tennessee Press, 1999).

64. Billings, "Religion as Opposition"; Billings and Samson, "Evangelical Christians and the Environment"; Smith, "Religion Shaping Mountain-top Removal Debate."

65. Catholic Bishops of Appalachia, *This Land Is Home to Me (1975) and At Home in the Web of Life (1995): Appalachian Pastoral Letters* (Martin, KY: Catholic Committee of Appalachia, 2007).

66. Catholic Committee of Appalachia, *Models of Ministry: An Evaluation* (Atlanta: Glenmary Research Center, 1989).

67. Michael Iafrate, "Decolonizing Appalachia: Postcolonial Theology in a U.S. American Region," *Catholicanarchy.org,* June 27, 2010, http://catholicanarchy.org/?p=1686.

68. Mary Ann Hinsdale, Helen M. Lewis, and S. Maxine Waller, *It Comes from the*

People: Community Development and Local Theology (Philadelphia: Temple University Press, 1995).

69. James M. Gustafson, *A Sense of the Divine: The Natural Environment from a Theocentric Perspective* (Cleveland, OH: Pilgrim Press, 1994), 62–66.

70. Throughout this book, I use the word "theocentric" to describe my approach. Despite the obvious affinities with Gustafson's theocentric ethics, I do not intend to refer to Gustafson's work unless I do so explicitly. It is simply the most apt word to describe Niebuhr's notion of God-centered interpretive frameworks, a notion that is at the foundation of my own approach. Some differences between Gustafson's approach and mine (and Niebuhr's) are noted in chapter 3.

1. Downstream Impacts

1. Jenkins, *Future of Ethics,* 67–69.

2. In my own research, the difficulty of finding anything approaching an unbiased treatment of MTR was quite frustrating, so I have endeavored to create such a treatment.

3. United States Environmental Protection Agency, *The Effects of Mountaintop Mines and Valley Fills on Aquatic Ecosystems of the Central Appalachian Coalfields (2011 Final)—EPA/600/R-09/138F* (Washington, DC: EPA, 2011), 16.

4. Ibid., 23.

5. Ibid., 24–26.

6. United States Environmental Protection Agency, *Mountaintop Mining/Valley Fills in Appalachia Draft Programmatic Environmental Impact Statement—EPA 903-R-00-013* (Philadelphia: EPA, 2003), III.D-4; Ken M. Fritz et al., "Structural and Functional Characteristics of Natural and Constructed Channels Draining a Reclaimed Mountaintop Removal and Valley Fill Coal Mine," *Journal of the North American Benthological Society* 29, no. 2 (April 6, 2010): 686, doi:10.1899/09-060.1.

7. US EPA, *Effects of Mountaintop Mines.*

8. Ibid., 29.

9. James Wickham et al., "The Overlooked Terrestrial Impacts of Mountaintop Mining," *BioScience* 63, no. 5 (May 1, 2013): 338, doi:10.1525/bio.2013.63.5.7.

10. US EPA, *Effects of Mountaintop Mines,* 1.

11. Ibid., 71–72.

12. Ibid., 86–87.

13. Fritz et al., "Structural and Functional Characteristics of Natural and Constructed Channels," 686.

14. US EPA, *Effects of Mountaintop Mines,* 88.

15. Fritz et al., "Structural and Functional Characteristics of Natural and Constructed Channels," 686.

16. Wickham et al., "Overlooked Terrestrial Impacts of Mountaintop Mining," 338.

17. Ibid., 339.

18. US EPA, *Mountaintop Mining/Valley Fills*, III.A-6.

19. Ibid., III.F-2.

20. Ibid., III.F-11.

21. Wickham et al., "Overlooked Terrestrial Impacts of Mountaintop Mining," 340.

22. US EPA, *Mountaintop Mining/Valley Fills*, III.F-14–15.

23. Ibid., III.B-11.

24. Ibid., III.F-6.

25. Ibid., III.F-9.

26. Ibid., III.F-7.

27. Ibid., III.F-1–2.

28. Wickham et al., "Overlooked Terrestrial Impacts of Mountaintop Mining," 343.

29. US EPA, *Mountaintop Mining/Valley Fills*, III.F-9.

30. Ibid., III.F-11.

31. Wickham et al., "Overlooked Terrestrial Impacts of Mountaintop Mining," 346.

32. Paul John Beggs, "Horizontal Cliffs: Mountaintop Mining and Climate Change," *Biodiversity and Conservation* 21, no. 14 (December 1, 2012): 3731–34, doi:10.1007/s10531-012-0387-y.

33. Ibid., 3733.

34. Wickham et al., "Overlooked Terrestrial Impacts of Mountaintop Mining," 342.

35. US EPA, *Mountaintop Mining/Valley Fills*, III.B-12.

36. Wickham et al., "Overlooked Terrestrial Impacts of Mountaintop Mining," 345.

37. Palmer et al., "Mountaintop Mining Consequences," 148.

38. David C. Holzman, "Mountaintop Removal Mining: Digging into Community Health Concerns," *Environmental Health Perspectives* 119, no. 11 (November 1, 2011): a480, doi:10.1289/ehp.119-a476.

39. Ibid.

40. Hitt and Hendryx, "Ecological Integrity of Streams."

41. Laura Esch and Michael Hendryx, "Chronic Cardiovascular Disease Mortality in Mountaintop Mining Areas of Central Appalachian States," *Journal of Rural Health* 27, no. 4 (September 2011): 350–57.

42. Melissa M. Ahern et al., "The Association between Mountaintop Mining and Birth Defects among Live Births in Central Appalachia, 1996–2003," *Environmental Research* 111, no. 6 (August 2011): 838–46, doi:10.1016/j.envres.2011.05.019.

43. Michael Hendryx and Melissa M. Ahern, "Mortality in Appalachian Coal Mining Regions: The Value of Statistical Life Lost," *Public Health Reports* 124, no. 4 (2009): 541–50.

44. Ibid., 548.

45. Shannon Elizabeth Bell and Richard York, "Community Economic Identity: The Coal Industry and Ideology Construction in West Virginia," *Rural Sociology* 75, no. 1 (January 2010): 113–15; Scott, *Removing Mountains*, 4.

46. Davis and Duffy, "King Coal vs. Reclamation," 677.

47. John R. Craynon, Emily A. Sarver, and David P. Robertson, "Could a Public Ecology Approach Help Resolve the Mountaintop Mining Controversy?" *Resources Policy* 38 (2013): 45.

48. Davis and Duffy, "King Coal vs. Reclamation," 677.

49. McIlmoil et al., "Coal and Renewables in Central Appalachia," 58. Similar studies have been conducted for other Appalachian states, with similar results. For this discussion, I focus on West Virginia because it accounts for approximately 44 percent of Appalachian coal production and is the state most dependent on the coal industry. Ibid., x.

50. Ibid., xii–xiii, 39–51.

51. Ibid., 58.

52. It should be noted that this is not what the study's authors are claiming. In fact, that is precisely my point: although the study illuminates the economic impact of coal on West Virginia's budget, it does not necessarily answer the question of whether the state would be better off without it.

53. Craynon, Sarver, and Robertson, "Could a Public Ecology Approach," 45.

54. Davis and Duffy, "King Coal vs. Reclamation," 678.

55. Ibid., 678–79.

56. Ibid., 679.

57. Ibid., 680.

58. Loeb, *Moving Mountains*, 92–93.

59. Davis and Duffy, "King Coal vs. Reclamation," 681.

60. US EPA, *Mountaintop Mining/Valley Fills*, I-8; Davis and Duffy, "King Coal vs. Reclamation," 684.

61. Davis and Duffy, "King Coal vs. Reclamation," 684–85.

62. Ibid., 686.

63. Craynon, Sarver, and Robertson, "Could a Public Ecology Approach," 45.

64. Ibid., 46.

65. Ibid., 45.

66. Ibid., 48.

67. Erik Reece, "The Power to Move Perceptions: Orwellian Language in the Land of Coal," in *Plundering Appalachia: The Tragedy of Mountaintop-Removal Coal Mining,* ed. Tom Butler and George Wuerthner, (San Rafael, CA: Earth Aware, 2009), 49.

68. Craynon, Sarver, and Robertson, "Could a Public Ecology Approach," 48.

69. David P. Robertson and R. Bruce Hull, "Public Ecology: An Environmental Science and Policy for Global Society," *Environmental Science & Policy* 6, no. 5 (October 2003): 399, doi:10.1016/S1462-9011(03)00077-7.

70. Craynon, Sarver, and Robertson, "Could a Public Ecology Approach," 46.

71. Robertson and Hull, "Public Ecology," 401.

72. Ibid., 399–410.

2. Environmental Ethics and the Construction of Values

1. By "ostensibly environmental," I mean to suggest that the ethical problem presented by MTR is, on the surface, one of environmental (that is, nonhuman) goods in conflict with other (economic, social, or political) goods. But as I show in this chapter, such neat categories are ultimately unhelpful and misleading.

2. Andrew Light and Holmes Rolston, "Introduction: Ethics and Environmental Ethics," in *Environmental Ethics: An Anthology,* ed. Andrew Light and Holmes Rolston, Blackwell Philosophy Anthologies 19 (Malden, MA: Blackwell, 2003), 8–9.

3. John O'Neill, "The Varieties of Intrinsic Value," in Light and Rolston, *Environmental Ethics: An Anthology,* 131–42.

4. Greta Gaard and Lori Gruen, "Ecofeminism: Toward Global Justice and Planetary Health," in Light and Rolston, *Environmental Ethics: An Anthology,* 277.

5. Ibid., 285.

6. Karen J. Warren and Jim Cheney, "Ecological Feminism and Ecosystem Ecology," in Light and Rolston, *Environmental Ethics: An Anthology,* 294–305.

7. Rosemary R. Ruether, *New Woman, New Earth: Sexist Ideologies and Human Liberation* (New York: Seabury Press, 1975); compare Gaard and Gruen, "Ecofeminism," 276.

8. Rosemary R. Ruether, *Gaia and God: An Ecofeminist Theology of Earth Healing* (San Francisco: HarperOne, 1994).

9. Ibid., 3, 200, 140.

10. Ibid., 87.

11. Ibid., 89.

12. Ibid., 56–57.

13. Ibid., 86, 142, 258.

14. Leonardo Boff, *Cry of the Earth, Cry of the Poor* (Maryknoll, NY: Orbis Books, 1997), 112–13.

15. Ibid., xi.

16. Ibid., 70–71.

17. The use of the masculine noun here is intentional.

18. Boff, *Cry of the Earth, Cry of the Poor,* 110–11.

19. Ibid., 15.

20. Ibid., 119–22.

21. Ibid., 112–13.

22. Ibid., 82–85.

23. Ibid., 114, chap. 11.

24. Ibid., 216.

25. K. S. Shrader-Frechette, *Environmental Justice: Creating Equality, Reclaiming Democracy* (Oxford: Oxford University Press, 2002), 6–12, 23.

26. Willis Jenkins, "After Lynn White: Religious Ethics and Environmental Problems," *Journal of Religious Ethics* 37, no. 2 (June 2009): 297.

27. Ibid.

28. Shrader-Frechette, *Environmental Justice.*

29. Jenkins, "After Lynn White," 298.

30. Loeb, *Moving Mountains;* Shnayerson, *Coal River;* Montrie, *To Save the Land and People.*

31. Morrone and Buckley, *Mountains of Injustice;* McNeil, *Combating Mountaintop Removal.*

32. Ben A. Minteer and Robert E. Manning, "Pragmatism in Environmental Ethics: Democracy, Pluralism, and the Management of Nature," in Light and Rolston, *Environmental Ethics: An Anthology,* 307–18; Anthony Weston, "Beyond Intrinsic Value: Pragmatism in Environmental Ethics," ibid., 307–18; Jenkins, "After Lynn White."

33. Thanks to Willis Jenkins for clarifying this point.

34. Weston, "Beyond Intrinsic Value," 311.

35. Ibid., 316.

36. Minteer and Manning, "Pragmatism in Environmental Ethics," 319.

37. Ibid., 325.

38. Anna L. Peterson, "Talking the Walk: A Practice-Based Environmental Ethic as Grounds for Hope," in *Ecospirit: Religions and Philosophies for the Earth,* ed. Laurel Kearns and Catherine Keller, Transdisciplinary Theological Colloquia (New York: Fordham University Press, 2007), 45–62.

39. Jenkins, "After Lynn White," 294.

40. Willis Jenkins, *Ecologies of Grace: Environmental Ethics and Christian Theology* (Oxford: Oxford University Press, 2008), 36–40.

41. Paul Robbins, *Political Ecology: A Critical Introduction,* Critical Introductions to Geography (Malden, MA: Blackwell, 2004), 13; compare Richard Peet and Michael Watts, eds., *Liberation Ecologies: Environment, Development, Social Movements* (London: Routledge, 1996), 6.

42. See, for example, Piers M. Blaikie and H. C. Brookfield, *Land Degradation and Society* (London: Methuen, 1987); Peet and Watts, *Liberation Ecologies.*

43. Robbins, *Political Ecology,* 13–14.

44. Ibid., 11.

45. Willis Jenkins, personal communication, August 3, 2012.

46. Peet and Watts, *Liberation Ecologies,* 13–27; Philip Anthony Stott and Sean Sullivan, *Political Ecology: Science, Myth and Power* (London: Arnold; Oxford University Press, 2000).

47. Catholic Bishops of Appalachia, *Appalachian Pastoral Letters,* 14–19.

48. Ibid., 19.

49. Ibid., 21.

50. Ibid., 44–45.

51. Ibid., 47, 84, 92.

52. Ibid., 42; Catholic Committee of Appalachia, *Models of Ministry,* 1–2.

53. Catholic Bishops of Appalachia, *Appalachian Pastoral Letters*, 21, 62.

54. Catholic Committee of Appalachia, *Models of Ministry;* Hinsdale, Lewis, and Waller, *It Comes from the People.*

55. Batteau, *Invention of Appalachia*, 13; Shapiro, *Appalachia on Our Mind*, 31; Hanna, "Representation and the Reproduction of Appalachian Space," 180–81.

56. Catholic Bishops of Appalachia, *Appalachian Pastoral Letters*, 36.

57. Batteau, *Invention of Appalachia*, 187.

58. Catholic Bishops of Appalachia, *Appalachian Pastoral Letters*, 32, 33.

59. Ibid., 34.

60. McNeil, *Combating Mountaintop Removal*, 2–7.

61. Ibid., 154–55.

62. Ibid., 168.

63. Ibid., 154, 168.

64. See, for example, Morrone and Buckley, *Mountains of Injustice.*

65. Jenkins, *Ecologies of Grace*, 37.

66. Melinda Bollar Wagner, "Space and Place, Land and Legacy," in *Culture, Environment, and Conservation in the Appalachian South,* ed. Benita J. Howell (Chicago: University of Illinois Press, 2002), 121–32.

67. Ibid., 130.

68. A second possible example is the Catholic Committee of Appalachia's *Models of Ministry,* which incorporates pragmatist methods (empirical research and experimentation) and values (participatory democracy, community organizing) to assess the church's efforts to respond to *This Land Is Home to Me.* However, because this report begins with the bishops' call for justice and liberation, it does not share pragmatism's central impulse—an antifoundationalist move away from abstract, metaethical debate.

69. Weston, "Beyond Intrinsic Value," 316.

70. I defend the utility of intuition and experience of place in the context of a theocentric approach. The key difference is that, for some pragmatists, experience seems to be primary and immediate, whereas in my approach, experience of nature, like all experience, is mediated by imaginations and interpretations and thus is subject to examination and critique from a theocentric perspective.

71. Mary Hufford, "Reclaiming the Commons: Narratives of Progress, Preservation, and Ginseng," in Howell, *Culture, Environment, and Conservation,* 100–120.

72. These conclusions bear clear similarities to McNeil's examination of grassroots groups discussed in the section on environmental justice. I discuss McNeil's work there, rather than here, because he addresses the lived experience of grassroots groups—a central emphasis of environmental justice—and McNeil himself argues that environmental justice is the most appropriate label for such groups.

73. Bell and York, "Community Economic Identity."

74. Robbins, *Political Ecology,* 13.

75. Arturo Escobar, "Constructing Nature: Elements for a Poststructural Political

Ecology," in Peet and Watts, *Liberation Ecologies*, 46–68.

76. Peet and Watts, *Liberation Ecologies*, 37; Hufford, "Reclaiming the Commons"; Robbins, *Political Ecology*, 56.

77. H. Richard Niebuhr, *Radical Monotheism and Western Culture: With Supplementary Essays*, Library of Theological Ethics (Louisville, KY: Westminster/John Knox Press, 1993), 34; compare Robert Merrihew Adams, *Finite and Infinite Goods: A Framework for Ethics* (New York: Oxford University Press, 1999), 232.

78. Niebuhr, *Radical Monotheism*, 100–101.

79. Ibid., 101–5.

80. Ibid., 104–5.

81. Ibid., 107.

82. Ibid., 111.

83. Ibid.

84. Ibid., 112.

85. Ibid., 33–34; Niebuhr, *Responsible Self*, 122–23.

86. Niebuhr describes this confused condition and how it is resolved by his ethic in *The Responsible Self*. I deal with that discussion more completely in the following chapter.

87. This is not to say that these tropes cannot be viewed as oppressive or distorted by a theocentric approach; indeed, I argue that, from a theocentric perspective, they may be described as evil not in themselves but insofar as they allow human values and purposes to stand in for the inclusive purposes of God.

88. Larry L. Rasmussen, *Earth Community, Earth Ethics* (Geneva: WCC Publications, 1996), 248, 261.

89. Ibid., 15, 328–35, 99–107.

90. Compare Gustafson, *Sense of the Divine*.

3. Relation, Revelation, and Revolution

1. H. Richard Niebuhr, *The Meaning of Revelation* (Louisville, KY: Westminster John Knox Press, 2006), 18; Niebuhr, *Radical Monotheism*, 110.

2. Niebuhr, *Meaning of Revelation*, 10.

3. James, "Responsibility Ethics and Postliberalism," 44.

4. Niebuhr, *Meaning of Revelation*, 19.

5. Naturally, this raises the question of what I mean by "the church," particularly because Christian groups that do not represent any institutional church are such central actors in the MTR debate. Following my interpretation of Niebuhr, I understand the church to refer to communities of persons that share a commitment (however feebly or imperfectly enacted) to the God of Jesus Christ as their organizing center of value; or, as Niebuhr puts it, "the church itself must be described . . . as the community which responds to God-in-Christ and Christ-in-God." H. Richard Niebuhr, "The Responsibility of the Church for Society," in *The Gospel, the Church and the World*, ed. Kenneth Scott Latourette (New York: Harper & Brothers, 1946), 117. I make no claim that the

church so understood is consistent in organizing its values this way; indeed, part of my goal (and part of Niebuhr's) is a critique of its failures to do so.

6. James, "Responsibility Ethics and Postliberalism," 46.

7. Niebuhr's radical monotheism may be more suited to the thoroughgoing monotheism of Islam, for example, than to Christianity.

8. William Schweiker, *Theological Ethics and Global Dynamics: In the Time of Many Worlds* (Malden, MA: Wiley-Blackwell, 2004), 18–19.

9. Douglas Ottati, introduction to Niebuhr, *Meaning of Revelation*, xxii.

10. Niebuhr, *Meaning of Revelation*, 11–12.

11. Niebuhr, *Responsible Self*, 1–2.

12. Niebuhr, *Radical Monotheism*, 107.

13. C. David Grant, *God the Center of Value: Value Theory in the Theology of H. Richard Niebuhr* (Fort Worth: Texas Christian University Press, 1984), 39.

14. Ibid., 40.

15. Niebuhr, *Radical Monotheism*, 107. Thus the word "value" can have two (not always distinguished) meanings: a general meaning, as the relationship whereby being confronts being, which can include relationships of positive or negative value; and, more narrowly, a relationship of positive value as opposed to disvalue.

16. Ibid., 29.

17. Niebuhr, *Responsible Self*, 48–54. Niebuhr's original locutions, "man-the-maker" and "man-the-citizen," have been adapted to be gender-inclusive.

18. Ibid., 55–68.

19. Ibid., 57.

20. Ibid., 61–63.

21. Ibid.

22. Ibid., 60.

23. Ibid., 112.

24. Ibid., 117–21, 142–43. Niebuhr actually states in *The Responsible Self* that faith can be positive or negative, "trust or distrust." The positive form—faith as trust—is more in keeping with his other works; see, for example, *Radical Monotheism*, 16, where faith is understood as confidence or trust and fidelity.

25. Niebuhr, *Radical Monotheism*, 110.

26. Ibid., 16.

27. Ibid., 24–31. See also chapter 2 of this volume.

28. Ibid., 38.

29. Niebuhr, *Meaning of Revelation*, 18.

30. Ibid., 73–76.

31. Ibid., 50.

32. Ibid., 72.

33. Ibid., 90.

34. Ibid., 80.

35. This transvaluing property of revelation calls to mind the words of a familiar hymn: "When I survey the wondrous cross, my every gain I count but loss."

36. Niebuhr, *Meaning of Revelation*, 15–16.

37. Niebuhr, *Radical Monotheism*, 38.

38. Grant, *God the Center of Value*, 68.

39. Niebuhr, *Responsible Self*, 149–54.

40. Niebuhr, *Meaning of Revelation*, 49–54.

41. Ibid., 53–54.

42. Ibid., 55–60.

43. Ibid., 95–98.

44. Ibid., 46. It is worth noting that in *The Responsible Self* and *Radical Monotheism* the relevant moral actor is most often the individual, while in *The Meaning of Revelation* it is the church. For Niebuhr, faith, or the decision to trust in the benevolence of God, seems to be a personal decision, whereas revelation and its transformation and illumination of history, which precipitate and sustain faith, are communal events. Niebuhr, *Responsible Self*, 120; Niebuhr, *Meaning of Revelation*, 28.

45. Niebuhr, *Responsible Self*, 177–78.

46. James M. Gustafson, *Ethics from a Theocentric Perspective*, vol. 1, *Theology and Ethics* (Chicago: University of Chicago Press, 1983), 18–19, 320.

47. James M. Gustafson, *Ethics from a Theocentric Perspective*, vol. 2, *Ethics and Theology* (Chicago: University of Chicago Press, 1992), 1.

48. Ibid., 2.

49. Ibid., 236.

50. Gustafson, *Ethics from a Theocentric Perspective*, 1:273–74.

51. Ibid., 166.

52. Gustafson, *Sense of the Divine*, 67.

53. Ibid., 72.

54. Ibid., 133.

55. Ibid., 132.

56. Ibid., 134.

57. Ibid., 135.

58. Ibid., 23.

59. Emilie M. Townes, *Womanist Ethics and the Cultural Production of Evil* (New York: Palgrave Macmillan, 2006), 5–8, 27.

60. Ibid., 11–24.

61. Ibid., 20.

62. Ibid., 27.

63. Ibid., 47–55.

64. Ibid., 61.

65. Ibid.

66. Ibid., 76.

67. Niebuhr, *Meaning of Revelation,* 46.

68. Ibid., 10. I am not suggesting that Townes fails to heed Niebuhr's cautions about historicity and particularity. She certainly does so; indeed, the particularity and the socially conditioned nature of knowledge are central tenets of her work. By saying that my starting point is "arguably more Niebuhrian," I simply mean that, with regard to the church, I am adopting the same particular perspective he did.

69. Ibid., 46.

70. Ibid.

71. Ibid., 99.

72. Ibid., xxxiv. The phrase is indeed "source evil," not "source *of* evil." I take this to mean that this act of absolutizing the relative is not only the origin of evil but evil itself.

73. Ibid., ix–x.

74. Stanley Hauerwas, "Why Christian Ethics Is Such a Bad Idea," in *Beyond Mere Health: Theology and Health Care in a Secular Society,* ed. Hilary D. Regan, Rodney B. Horsfield, and Gabrielle L. McMullen (Kew Victoria, Australia: Australian Theological Forum, 1996), 74–75; John Howard Yoder, *The Politics of Jesus* (Grand Rapids, MI: Eerdmans, 1994), 17–18, 101–2.

75. Niebuhr, *Meaning of Revelation,* xxxiv, 80.

76. Ibid., 78.

77. Ibid., 58–69.

78. Ibid., 98–99; Niebuhr, *Responsible Self,* 175–78.

79. Stanley Hauerwas, *After Christendom? How the Church Is to Behave if Freedom, Justice, and a Christian Nation Are Bad Ideas* (Nashville, TN: Abingdon Press, 1991), 168; Rasmussen, *Earth Community, Earth Ethics,* 248.

80. Niebuhr, *Responsible Self,* 67.

81. Ibid., 60.

4. The Meanings of the Mountains

1. Batteau, *Invention of Appalachia,* 13.

2. Hanna, "Representation and Reproduction of Appalachian Space."

3. Ibid., 187.

4. Ibid., 186.

5. Niebuhr, *Meaning of Revelation,* 54–55.

6. For a paradigmatic and extremely influential (for better or worse) example, see Caudill, *Night Comes to the Cumberlands.* For documentation of this story in the history of one mining town, see Gaventa, *Power and Powerlessness.* See also Grace Edwards, *A Handbook to Appalachia: An Introduction to the Region* (Knoxville: University of Tennessee Press, 2006), 1–26, 51–84; Chuck Shuford, "What Happens When You Don't Own the Land | Daily Yonder | Keep It Rural," July 3, 2009, http://www.dailyyonder .com/what-happens-when-you-dont-own-land/2009/07/03/2205. Compare Barbara Ellen Smith, "Legends of the Fall: Contesting Economic History," in *Christianity in*

Appalachia: Profiles in Regional Pluralism, ed. Bill Leonard (Knoxville: University of Tennessee Press, 1999), 1–17.

7. Gaventa, *Power and Powerlessness*, 80.

8. Ibid., 14–20.

9. Ibid., 13.

10. Ibid., 63–68.

11. This role of religion, and particularly the role of mountain churches such as the Pentecostal and Holiness churches, in encouraging quiescence and dependence continues to be invoked in discussions of religion in Appalachia and in the debate over MTR. Of course, such a straightforward view of the church's complicity, or even the complicity of these particular churches, is oversimplified. See Corinne Almquist, "I Have Been to the Mountaintop, but It Wasn't There: Christian Responses to Mountaintop Removal Coal Mining in Appalachia" (senior thesis, Middlebury College, 2009); Billings and Samson, "Evangelical Christians and the Environment"; David Corbin, *Life, Work, and Rebellion in the Coal Fields: The Southern West Virginia Miners, 1880–1922* (Urbana: University of Illinois Press, 1981), 146–75.

12. Gaventa, *Power and Powerlessness*, 93.

13. Ibid., 40–43.

14. Ibid., 252–61.

15. Burns, *Bringing Down the Mountains*, 2; Morrone and Buckley, *Mountains of Injustice*, xii; Jason Howard, "Appalachia Turns on Itself," *New York Times*, July 8, 2012, Opinion sec.; Silas House, "My Polluted Kentucky Home," *New York Times*, February 19, 2011, Opinion sec.

16. Burns, *Bringing Down the Mountains*, 1–8; Stephen J. Scanlan, "The Theoretical Roots and Sociology of Environmental Justice in Appalachia," in *Mountains of Injustice: Social and Environmental Justice in Appalachia*, ed. Michele Morrone and Geoffrey L. Buckley (Athens: Ohio University Press, 2011), 16; Howard, "Appalachia Turns on Itself."

17. A "broad form deed" was a complicated document that severed mineral rights from surface rights, granting the purchaser the mineral rights to a property as well as the right to alter its surface to extract the minerals. Ronald D. Eller, *Uneven Ground: Appalachia since 1945* (Lexington: University Press of Kentucky, 2008), 37–38.

18. Scanlan, "Theoretical Roots and Sociology of Environmental Justice," 22.

19. Quoted in Almquist, "I Have Been to the Mountaintop," 152.

20. Onleilove Alston, "Destroying West Virginia, One Mountain at a Time: Christians Battle King Coal to Save Appalachia," *Sojourners* 39, no. 6 (June 2010): 18–20.

21. Josh MacIvor-Andersen, "A Brief History of Coal," *Prism: America's Alternative Evangelical Voice* 14, no. 6 (December 2007): 14.

22. Catholic Bishops of Appalachia, *Appalachian Pastoral Letters*, 14.

23. Compare Smith, "Legends of the Fall."

24. Catholic Bishops of Appalachia, *Appalachian Pastoral Letters*, 59–60.

25. Ibid., 33.

26. Scott, *Removing Mountains,* 13–14.

27. Ibid., 12; Smith, "Legends of the Fall," 7.

28. Compare Fisher, *Fighting Back in Appalachia,* 4, 235–336.

29. Unitarian Universalist Association, "End Mountaintop Removal Coal Mining: 2006 Action of Immediate Witness," http://ilovemountains.org/resolutions#uua. I acknowledge that the Unitarian Universalist Association is not, strictly speaking, a Christian organization; I include its statement here because it most clearly expresses the view of power that is operative in the other statements and because it is sufficiently similar that I believe it can be considered representative.

30. Gaventa, *Power and Powerlessness,* 14–15.

31. "iLoveMountains.org—Resolutions of Faith—End Mountaintop Removal Coal Mining," http://ilovemountains.org/resolutions#ec; West Virginia Council of Churches, "Statement on Mountaintop Removal" (Charleston: West Virginia Council of Churches, September 11, 2007).

32. Quoted in Almquist, "I Have Been to the Mountaintop," 99.

33. One noteworthy example is the "authentic local theology" of theologian Mary Ann Hinsdale and activist Helen Matthews Lewis, who argue that communities themselves should be a prime locus for political theology. At the same time, their work seems to assume a univocal community that is problematic. Hinsdale, Lewis, and Waller, *It Comes from the People.*

34. Burns, *Bringing Down the Mountains,* 5.

35. Shnayerson, *Coal River,* 19–20.

36. Reece, *Lost Mountain,* 62–63; Loeb, *Moving Mountains,* 151–52; Scott, *Removing Mountains,* 211; Bill Raney, "Bill Raney: Congress Must Rein in an Arrogant EPA," December 31, 2010, http://dailymail.com/Opinion/Commentary/201012301192?page=1&build=cache.

37. Troy Andes cited in Ken Ward, "Massey Protest Update: 14 Arrested; Accusations Fly," June 18, 2009, http://blogs.wvgazette.com/coaltattoo/2009/06/18/massey-protest-update-14-arrested-accusations-fly/.

38. Scott, *Removing Mountains.*

39. Batteau, *Invention of Appalachia,* 33.

40. Ibid., 29, 33.

41. Ibid., 37.

42. Shapiro, *Appalachia on Our Mind,* 32–58.

43. Ibid., 85–87.

44. Scott, *Removing Mountains,* 31.

45. Ibid., 63.

46. Ibid., 113.

47. Ibid., 109, 173–74.

48. Ibid., 210.

49. Ibid., 216.

50. Ibid., 179.

51. Scanlan, "Theoretical Roots and Sociology of Environmental Justice," 14.

52. Scott, *Removing Mountains,* 129.

53. Ibid., 84.

54. Ibid., 33–34.

55. Batteau, *Invention of Appalachia,* 186–93.

56. Ibid., 186.

57. Ibid., 186–87.

58. Scott, *Removing Mountains,* 218–19.

59. Batteau, *Invention of Appalachia,* 187.

60. Ibid.

61. Unitarian Universalist Association, "End Mountaintop Removal Coal Mining."

62. United Methodist Church, "Cease Mountaintop Removal Coal Mining," 2000, http://ilovemountains.org/resolutions.

63. Presbyterian Church (USA) 217th General Assembly, "Commissioners' Resolution," http://ilovemountains.org/resolutions (accessed August 1, 2012).

64. Lon Oliver, personal communication, July 28, 2010.

65. Catholic Bishops of Appalachia, *Appalachian Pastoral Letters,* 21–22, 48, 55–59.

66. Ibid., 61–63.

67. Batteau, *Invention of Appalachia,* 186–87.

68. Carol Warren, interview with the author, Charleston, WV, August 9, 2010.

69. Father John Rausch, interview with the author, Stanton, KY, August 10, 2010.

70. Father Stan Holmes, interview with the author, Charleston, WV, August 9, 2010.

71. Denise Giardina, "Keynote Address, Christians for the Mountains Conference," November 2005, http://christiansforthemountains.org/site/Topics/About/deniseGiardina.html.

72. Butler and Wuerthner, *Plundering Appalachia.*

73. Ibid., 59.

74. Wendell Berry, foreword to *Lost Mountain: A Year in the Vanishing Wilderness: Radical Strip Mining and the Devastation of Appalachia,* by Erik Reece (New York: Riverhead Books, 2006), xvi.

75. Scott, *Removing Mountains,* 96, 174.

76. Almquist, "I Have Been to the Mountaintop," 162.

77. Ibid., 172.

78. Giardina, "Keynote Address."

79. "iLoveMountains.org—Resolutions of Faith—End Mountaintop Removal Coal Mining." Undoubtedly, terms like "demolition" and "removal" have their own discursive baggage. They are, however, arguably more precise descriptions of the process of MTR and, as will become clear, do not involve the broad implications of "destruction."

80. D. L. Johnson and L. A. Lewis cited in Robbins, *Political Ecology,* 91.

81. Ibid.

82. Ibid., 94.

83. Ibid., 97.

84. Ibid., 90.

85. Presbyterian Church (USA) 217th General Assembly, "Commissioners' Resolution," emphasis added.

86. West Virginia Coal Association, *West Virginia Coal; Reclaiming the Future*. I have personally visited a reclaimed mine site, and it was virtually indistinguishable from the surrounding forested hillside.

87. Burns, *Bringing Down the Mountains*, 118–40; Butler and Wuerthner, *Plundering Appalachia*, 49; Scott, *Removing Mountains*, 95. For many MTR opponents, the fact that "most" or "almost all" reclamation is done so carelessly justifies the conclusion that real reclamation is illusory. For example, see Burns's claim that successful reclamation is impossible because current regulations are inadequate (*Bringing Down the Mountains*, 123).

88. Scott, *Removing Mountains*, 176–79.

89. Ibid., 83.

90. Reece, *Lost Mountain*, 142.

91. Robbins, *Political Ecology*, 101–3.

92. Ibid.

93. *Reclaiming the Future*.

94. Hufford, "Reclaiming the Commons," 113.

95. Ibid.; Scott, *Removing Mountains*, 117–21, 211.

96. Scott, *Removing Mountains*, 123.

97. Ibid., 131–32, 178.

98. Robbins, *Political Ecology*, 102.

99. Thomas R. Shannon, "The Economy of Appalachia," in Edwards, *Handbook to Appalachia*, 71.

100. Robbins, *Political Ecology*, 103–5.

101. Ibid., xvii, 105, 208–9.

102. Even for those who advocate a complete stop to MTR, this process of clarification is necessary for the better reclamation of existing mine sites.

103. Smith, "Legends of the Fall," 13.

104. Scott, *Removing Mountains*, 27.

5. All My Holy Mountain

1. Michel Foucault, *Power/Knowledge: Selected Interviews and Other Writings, 1972-1977*, ed. Colin Gordon (New York: Vintage, 1980), 78–92, 131–33.

2. Niebuhr, *Meaning of Revelation*, 60, 66.

3. Niebuhr, *Responsible Self*, 117–21; see also chapter 3.

4. Ibid., 119; Niebuhr, *Meaning of Revelation*, 97–99.

5. Niebuhr, *Responsible Self,* 126.

6. Ibid., 166.

7. Ibid., 164–66.

8. Ibid., 177; Niebuhr, *Meaning of Revelation,* 97.

9. Niebuhr, *Meaning of Revelation,* 98.

10. Fisher, *Fighting Back in Appalachia,* 11–12.

11. Scott, *Removing Mountains,* 12, 221.

12. Hanna, "Representation and Reproduction of Appalachian Space," 186.

13. Compare Iafrate, "Decolonizing Appalachia."

14. Catholic Bishops of Appalachia, *Appalachian Pastoral Letters,* 33.

15. Ibid., 98.

16. Lon Oliver, interview with the author, Berea, KY, August 10, 2010.

17. Niebuhr, *Meaning of Revelation,* 99.

18. Gustafson, *Sense of the Divine,* 73–74.

19. Niebuhr, *Responsible Self,* 142.

20. Oliver interview.

21. ABC News, "A Hidden America: Children of the Mountains," February 10, 2009, http://abcnews.go.com/2020/story?id=6845770&page=1.

22. Oliver interview.

23. Allen Johnson, interview with the author, Marlinton, WV, August 6, 2010.

24. Niebuhr, *Responsible Self,* 71–84.

25. Ibid., 121–25.

26. Niebuhr, *Radical Monotheism,* 103–6.

27. Ibid., 31–34, 112–13.

28. Of course, this implication can be viewed in the opposite direction: what we value, we construct as similar; what we do not value, we construct as other.

29. Scott, *Removing Mountains,* 62–63.

30. Troy Andes cited in Ward, "Massey Protest Update."

31. Niebuhr, *Radical Monotheism,* 113.

32. Niebuhr, *Meaning of Revelation,* 64–65.

33. Smith, "Legends of the Fall," 13.

34. Niebuhr, *Meaning of Revelation,* xxxiv.

35. Scott, *Removing Mountains,* 7.

36. Niebuhr, *Responsible Self,* 56–57.

37. Batteau, *Invention of Appalachia,* 13.

38. Hanna, "Representation and Reproduction of Appalachian Space," 186–87.

39. Scott, *Removing Mountains,* 212–14.

40. Niebuhr, *Meaning of Revelation,* 80.

41. Ibid., 64–65.

42. Ibid., 64.

43. Batteau, *Invention of Appalachia,* 186–87.

44. Niebuhr, *Radical Monotheism*, 102.

45. Reverend Stan Holmes, interview with author, Charleston, WV, August 9, 2010.

46. Andrew Jordon, interview with author, Kanawha County, WV, August 9, 2010.

47. Billings and Samson, "Evangelical Christians and the Environment," 5.

48. Niebuhr, *Meaning of Revelation*, 63–64.

49. Loyal Jones, "Mountain Religion: An Overview," in *Christianity in Appalachia: Profiles in Regional Pluralism,* ed. Bill Leonard (Knoxville: University of Tennessee Press, 1999), 92–98.

50. Ibid., 97.

51. Deborah Vansau McCauley, "Mountain Holiness," in Leonard, *Christianity in Appalachia,* 104.

52. Billings and Samson, "Evangelical Christians and the Environment," 5.

53. Ibid.

54. Niebuhr, *Responsible Self,* 71–84.

55. Johnson interview.

56. Scott, *Removing Mountains,* 102.

57. Lon Oliver, personal communication, July 28, 2010.

58. Jordon interview.

59. Johnson interview.

60. Niebuhr, *Responsible Self,* 126.

61. Ibid., 173.

62. Niebuhr, *Meaning of Revelation,* 15–19.

63. Gustafson, *Sense of the Divine,* 72–73.

64. Ibid., 66, 72.

65. Ibid., 72. Gustafson notes that there may be some exceptional cases, such as the eradication of smallpox, in which the benefits are overwhelming and the negative consequences are negligible. Some might argue that MTR is as clear-cut as this, but I do not believe this to be the case. In the final chapter, however, I raise a strong theocentric challenge to MTR from a different angle.

66. Ibid., 13.

67. Jordon interview.

68. Oliver interview.

69. Samir Doshi, "Restoration Economy: Reclaiming the Land and Our Communities," in *Plundering Appalachia: The Tragedy of Mountaintop-Removal Coal Mining,* ed. Tom Butler and George Wuerthner (San Rafael, CA: Earth Aware, 2009), 61.

70. Father John S. Rausch, "Sowing My Community Back," *Steubenville (OH) Register,* January 31, 2003.

71. Carol Warren, interview with the author, Charleston, WV, August 9, 2010; Daniel McGlynn, "Move Not Those Bones," *Sierra* 97, no. 2 (April 2012): 28–33.

72. This recurrence indicates how seriously I take this criticism.

73. Niebuhr, *Responsible Self,* 173.

6. Loving the Mountains

1. Niebuhr, *Meaning of Revelation*, 69; Niebuhr, *Responsible Self*, 143–44.

2. Niebuhr, *Radical Monotheism*, 107.

3. Niebuhr, *Meaning of Revelation*, 46.

4. Gustafson, *Sense of the Divine*, 13.

5. Compare Foucault, *Power/Knowledge*, 78–92.

6. H. Richard Niebuhr, *The Purpose of the Church and Its Ministry: Reflections on the Aims of Theological Education* (New York: Harper, 1956), 129.

7. James M. Gustafson, "Christian Ethics and Social Policy," in *Faith and Ethics: The Theology of H. Richard Niebuhr*, ed. Paul Ramsey (New York: Harper, 1957), 134.

8. Ibid., 133.

9. Ibid.

10. Gustafson, *Sense of the Divine*, 48.

11. Ibid., 72.

12. Niebuhr, *Meaning of Revelation*, 46.

13. Ibid., 64.

14. Niebuhr, "Responsibility of the Church for Society," 128.

15. Ibid., 131.

16. Allen Johnson, interview with the author, August 6, 2010.

17. James M. Gustafson, introduction to *The Responsible Self: An Essay in Christian Moral Philosophy*, by H. Richard Niebuhr (New York: Harper & Row, 1963), 18.

18. Niebuhr, *Meaning of Revelation*, 44.

19. Gustafson, introduction to *Responsible Self*, 32.

20. Lon Oliver, interview with the author, August 10, 2010.

21. Niebuhr, *Meaning of Revelation*, 49–50.

22. Gustafson, "Christian Ethics and Social Policy," 127.

23. Niebuhr, *Meaning of Revelation*, 70–72.

24. Gustafson, "Christian Ethics and Social Policy," 127.

25. Niebuhr, *Meaning of Revelation*, 96–98.

26. John M. Glen, "Like a Flower Slowly Blooming: Highlander and the Nurturing of an Appalachian Movement," in *Fighting Back in Appalachia: Traditions of Resistance and Change*, ed. Stephen L. Fisher (Philadelphia: Temple University Press, 1993), 31–55.

27. Hinsdale, Lewis, and Waller, *It Comes from the People*.

28. Allen Johnson, personal communication, February 13, 2010; Johnson interview; Walter Wink, *The Powers that Be: Theology for a New Millennium* (New York: Doubleday, 1998).

29. Gustafson, "Christian Ethics and Social Policy," 131.

30. Niebuhr, *Radical Monotheism*, 47–48.

31. Gustafson, "Christian Ethics and Social Policy," 128.

32. Niebuhr, *Responsible Self*, 59–60.

33. Niebuhr, *Meaning of Revelation*, 72.

34. Gustafson, "Christian Ethics and Social Policy," 134.

35. Roy Gene Crist, personal communication, July 20, 2010.

36. Smith, "Religion Shaping Mountain-top Removal Debate."

37. Niebuhr, *Responsible Self*, 94–98; Niebuhr, *Meaning of Revelation*, 49–50.

38. Niebuhr, *Meaning of Revelation*, 80–81.

39. Ibid., 72.

40. Gustafson, "Christian Ethics and Social Policy," 134.

41. John Rausch, interview with the author, August 10, 2010; Catholic Bishops of Appalachia, *Appalachian Pastoral Letters.*

42. Catholic Committee of Appalachia, *Models of Ministry.*

43. Niebuhr, *Responsible Self*, 118–21.

44. Ibid., 124–25, emphasis in original.

45. Niebuhr, *Meaning of Revelation*, 19, 80–81.

46. Ibid., 19.

47. Ibid., 72.

48. Niebuhr, *Responsible Self*, 172.

49. Niebuhr, *Meaning of Revelation*, xxxiv.

50. Ibid., 18.

51. Ibid., 19.

52. That a theocentric perspective would undermine prophetic resistance to ecological violence is ultimately not surprising in light of Niebuhr's view of the violence and suffering of war as a call to repentance, rather than resistance. See H. Richard Niebuhr, "War as the Judgment of God," in *War as Crucifixion: Essays on Peace, Violence and Just War from the Christian Century,* ed. John M. Buchanan and David Heim (Chicago: Christian Century Press, 2002), 17–23.

53. Niebuhr, *Responsible Self*, 173.

54. Ibid.

55. Niebuhr, *Meaning of Revelation*, 51–52.

56. Ibid., 54–55.

57. Gustafson, *Sense of the Divine*, 17.

58. Ibid., 26–27.

59. I am not suggesting (nor is Gustafson) that value is perceived only negatively; I am (and he is) simply arguing that this perception of disvalue seems clearer and more immediate than the recognition of value.

60. Niebuhr, *Radical Monotheism*, 36–37.

61. Gustafson, *Sense of the Divine*, 5.

62. I am grateful to Willis Jenkins for suggesting this approach to the issue.

63. Iris Murdoch, *The Sovereignty of Good,* 2nd ed. (New York: Routledge, 2001), 37.

64. Ibid., 41.

65. Ibid., 89.

66. Ibid., 83.

67. Gustafson, *Sense of the Divine,* 23.

68. Murdoch, *Sovereignty of Good,* 86–89.

69. Indeed, there are many underground miners who actively oppose MTR based on their intimate experience of the mountains. However, it seems to me that the mechanized practices and enormous scale of MTR do not allow the kind of attention Murdoch describes. Whether smaller-scale surface mining might be a way of actively loving the mountains is a question that perhaps only a miner can answer definitively.

70. Berry is a fierce opponent of MTR, and he clearly implies that a "Christian strip mine" is a contradiction in terms.

71. Wendell Berry, *The Art of the Commonplace: The Agrarian Essays of Wendell Berry,* ed. Norman Wirzba (Emeryville, CA: Shoemaker & Hoard, 2002), 22.

72. Ibid., 299.

73. Ibid., 27, emphasis in original.

74. P. J. Crutzen and E. Stoermer, "The Anthropocene," *Global Change Newsletter* 41, no. 1 (2000); Jenkins, *Future of Ethics,* 1–3.

75. Crutzen and Stoermer, "The Anthropocene."

76. Jenkins, *Future of Ethics,* 2.

77. Niebuhr, "War as the Judgment of God."

78. Whether such reckless practices are the exception or the rule depends on whom one asks. In any case, such broad characterizations are less useful than the careful description of better and worse ways of mining and the encouragement, through policy and cooperation, of the former.

79. McIlmoil et al., "Coal and Renewables in Central Appalachia."

80. Niebuhr, *Meaning of Revelation,* 91.

81. Gustafson, *Ethics from a Theocentric Perspective,* 2:299.

82. Oliver interview.

83. Stan Holmes, interview with the author, Charleston, WV, August 9, 2010; Crist, personal communication.

84. Niebuhr, *Radical Monotheism,* 31.

85. Compare Niebuhr, "Responsibility of the Church for Society," 129.

Bibliography

ABC News. "A Hidden America: Children of the Mountains," February 10, 2009. http://abcnews.go.com/2020/story?id=6845770&page=1.

Adams, Robert Merrihew. *Finite and Infinite Goods: A Framework for Ethics.* New York: Oxford University Press, 1999.

Ahern, Melissa M., Michael Hendryx, Jamison Conley, Evan Fedorko, Alan Ducatman, and Keith J. Zullig. "The Association between Mountaintop Mining and Birth Defects among Live Births in Central Appalachia, 1996–2003." *Environmental Research* 111, no. 6 (August 2011): 838–46.

Almquist, Corinne. "I Have Been to the Mountaintop, but It Wasn't There: Christian Responses to Mountaintop Removal Coal Mining in Appalachia." Senior thesis, Middlebury College, 2009.

Alston, Onleilove. "Destroying West Virginia, One Mountain at a Time: Christians Battle King Coal to Save Appalachia." *Sojourners* 39, no. 6 (June 2010): 18–20.

Appalachian Voices. "Update: Extent of Mountaintop Mining in Appalachia as of 2008." http://ilovemountains.org/reclamation-fail/details.php#extent_study_2012.

Associated Press. "W.Va. Churches Slam Proposed Mining Rule." *New York Times,* October 11, 2007. http://www.wvcc.org/docs/InTheNews/newyorktimes.html.

Batteau, Allen. *The Invention of Appalachia: The Anthropology of Form and Meaning.* Tucson: University of Arizona Press, 1990.

Beggs, Paul John. "Horizontal Cliffs: Mountaintop Mining and Climate Change." *Biodiversity and Conservation* 21, no. 14 (December 1, 2012): 3731–34.

Bell, Shannon Elizabeth, and Richard York. "Community Economic Identity: The Coal Industry and Ideology Construction in West Virginia." *Rural Sociology* 75, no. 1 (January 2010): 111–43.

Berry, Wendell. *The Art of the Commonplace: The Agrarian Essays of Wendell Berry.* Edited by Norman Wirzba. Emeryville, CA: Shoemaker & Hoard, 2002.

———. Foreword to *Lost Mountain: A Year in the Vanishing Wilderness: Radical Strip Mining and the Devastation of Appalachia,* by Erik Reece, xv–xvii. New York: Riverhead Books, 2006.

Biggers, Jeff. "Updated: VIDEO: Nonviolent Goldman Prize Winner Attacked by Massey Supporter: 94-Year-Old Hechler, Hannah, Hansen Arrested at Coal River." *Huffington Post.* http://www.huffingtonpost.com/jeff-biggers/live-at-coal-river-daryl_b_219628.html.

Billings, Dwight B. "Religion as Opposition: A Gramscian Analysis." *American Journal of Sociology* 96, no. 1 (July 1990): 1–31.

Billings, Dwight B., and Will Samson. "Evangelical Christians and the Environment: 'Christians for the Mountains' and the Appalachian Movement against Mountaintop Removal Coal Mining." *Worldviews: Global Religions, Culture & Ecology* 16, no. 1 (2012): 1–29.

Billings, Dwight B., and Shaunna L. Scott. "Religion and Political Legitimation." *Annual Review of Sociology* 20, no. 1 (1994): 173–201.

Blaikie, Piers M., and H. C. Brookfield. *Land Degradation and Society.* London: Methuen, 1987.

Boff, Leonardo. *Cry of the Earth, Cry of the Poor.* Maryknoll, NY: Orbis Books, 1997.

Buchanan, John M., and David Heim, eds. *War as Crucifixion: Essays on Peace, Violence and Just War from the Christian Century.* Chicago: Christian Century Press, 2002.

Burns, Shirley Stewart. *Bringing Down the Mountains: The Impact of Mountaintop Removal on Southern West Virginia Communities.* Morgantown: West Virginia University Press, 2007.

Butler, Tom, and George Wuerthner, eds. *Plundering Appalachia: The Tragedy of Mountaintop-Removal Coal Mining.* San Rafael, CA: Earth Aware, 2009.

Callahan, Richard. *Work and Faith in the Kentucky Coal Fields: Subject to Dust.* Bloomington: Indiana University Press, 2009.

Casey, Edward S. *Getting Back into Place: Toward a Renewed Understanding of the Place-World.* Bloomington: Indiana University Press, 2009.

Catholic Bishops of Appalachia. *This Land Is Home to Me (1975) and At Home in the Web of Life (1995): Appalachian Pastoral Letters.* Martin, KY: Catholic Committee of Appalachia, 2007.

Catholic Committee of Appalachia. *Models of Ministry: An Evaluation.* Atlanta: Glenmary Research Center, 1989.

Caudill, Harry M. *Night Comes to the Cumberlands: A Biography of a Depressed Area.* Boston: Little, Brown, 1963.

Corbin, David. *Life, Work, and Rebellion in the Coal Fields: The Southern West Virginia Miners, 1880–1922.* Urbana: University of Illinois Press, 1981.

Craynon, John R., Emily A. Sarver, and David P. Robertson. "Could a Public Ecology Approach Help Resolve the Mountaintop Mining Controversy?" *Resources Policy* 38 (2013): 44–49.

Crutzen, P. J., and E. Stoermer. "The Anthropocene." *Global Change Newsletter* 41, no. 1 (2000).

Davis, Charles E., and Robert J. Duffy. "King Coal vs. Reclamation: Federal Regulation of Mountaintop Removal Mining in Appalachia." *Administration & Society* 41, no. 6 (October 1, 2009): 674–92.

Doshi, Samir. "Restoration Economy: Reclaiming the Land and Our Communities." In *Plundering Appalachia: The Tragedy of Mountaintop-Removal Coal Mining,* edited by Tom Butler and George Wuerthner, 61. San Rafael, CA: Earth Aware, 2009.

Eckholm, Erik. "West Virginia Sues over Mountaintop Mining Limits,"

October 13, 2010. http://green.blogs.nytimes.com/2010/10/06/west-virginia-sues-over-mountaintop-mining-limits/?hp.

Edwards, Grace. *A Handbook to Appalachia: An Introduction to the Region.* Knoxville: University of Tennessee Press, 2006.

Eller, Ronald D. *Uneven Ground: Appalachia since 1945.* Lexington: University Press of Kentucky, 2008.

Esch, Laura, and Michael Hendryx. "Chronic Cardiovascular Disease Mortality in Mountaintop Mining Areas of Central Appalachian States." *Journal of Rural Health* 27, no. 4 (September 2011): 350–57.

Escobar, Arturo. "Constructing Nature: Elements for a Poststructural Political Ecology." In *Liberation Ecologies: Environment, Development, Social Movements,* edited by Richard Peet and Michael Watts, 46–68. London: Routledge, 1996.

Fisher, Stephen L., ed. *Fighting Back in Appalachia: Traditions of Resistance and Change.* Philadelphia: Temple University Press, 1993.

Foucault, Michel. *Power/Knowledge: Selected Interviews and Other Writings, 1972–1977.* Edited by Colin Gordon. New York: Vintage, 1980.

Fritz, Ken M., Stephanie Fulton, Brent R. Johnson, Chris D. Barton, Jeff D. Jack, David A. Word, and Roger A. Burke. "Structural and Functional Characteristics of Natural and Constructed Channels Draining a Reclaimed Mountaintop Removal and Valley Fill Coal Mine." *Journal of the North American Benthological Society* 29, no. 2 (April 6, 2010): 673–89.

Gaard, Greta, and Lori Gruen. "Ecofeminism: Toward Global Justice and Planetary Health." In *Environmental Ethics: An Anthology,* edited by Andrew Light and Holmes Rolston. Blackwell Philosophy Anthologies 19. Malden, MA: Blackwell, 2003.

Gaventa, John. *Power and Powerlessness: Quiescence and Rebellion in an Appalachian Valley.* Urbana: University of Illinois Press, 1980.

Geller, Phylis. *Coal Country.* Evening Star, 2009. Film.

Giardina, Denise. "Keynote Address, Christians for the Mountains Conference," November 2005. http://christiansforthemountains.org/site/Topics/About/deniseGiardina.html.

Glen, John M. "Like a Flower Slowly Blooming: Highlander and the Nurturing of an Appalachian Movement." In *Fighting Back in Appalachia: Traditions of Resistance and Change,* edited by Stephen L. Fisher, 31–55. Philadelphia: Temple University Press, 1993.

Goodell, Jeff. *Big Coal: The Dirty Secret behind America's Energy Future.* Boston: Houghton Mifflin, 2006.

Grant, C. David. *God the Center of Value: Value Theory in the Theology of H. Richard Niebuhr.* Fort Worth: Texas Christian University Press, 1984.

Gustafson, James M. "Christian Ethics and Social Policy." In *Faith and Ethics: The Theology of H. Richard Niebuhr,* edited by Paul Ramsey, 119–39. New York: Harper, 1957.

———. *Ethics from a Theocentric Perspective.* Vol. 1, *Theology and Ethics.* Chicago: University of Chicago Press, 1983.

———. *Ethics from a Theocentric Perspective.* Vol. 2, *Ethics and Theology.* Chicago: University of Chicago Press, 1992.

———. Introduction to *The Responsible Self: An Essay in Christian Moral Philosophy,* by H. Richard Niebuhr. New York: Harper & Row, 1963.

———. *A Sense of the Divine: The Natural Environment from a Theocentric Perspective.* Cleveland, OH: Pilgrim Press, 1994.

Hanna, S. P. "Representation and the Reproduction of Appalachian Space: A History of Contested Signs and Meanings." *Historical Geography* 28 (2000): 179–207.

Hauerwas, Stanley. *After Christendom? How the Church Is to Behave if Freedom, Justice, and a Christian Nation Are Bad Ideas.* Nashville, TN: Abingdon Press, 1991.

———. "Why Christian Ethics Is Such a Bad Idea." In *Beyond Mere Health: Theology and Health Care in a Secular Society,* edited by Hilary D. Regan, Rodney B. Horsfield, and Gabrielle L. McMullen, 64–79. Kew Victoria, Australia: Australian Theological Forum, 1996.

Hendryx, Michael, and Melissa M. Ahern. "Mortality in Appalachian Coal Mining Regions: The Value of Statistical Life Lost." *Public Health Reports* 124, no. 4 (2009): 541–50.

Hendryx, Michael, and Keith J. Zullig. "Higher Coronary Heart Disease and Heart Attack Morbidity in Appalachian Coal Mining Regions." *Preventive Medicine* 49, no. 5 (November 2009): 355–59. doi:10.1016/j.ypmed.2009.09.011.

Hinsdale, Mary Ann, Helen M. Lewis, and S. Maxine Waller. *It Comes from the People: Community Development and Local Theology.* Philadelphia: Temple University Press, 1995.

Hitt, Nathaniel P., and Michael Hendryx. "Ecological Integrity of Streams Related to Human Cancer Mortality Rates." *EcoHealth* 7, no. 1 (April 2010): 91–104.

Holzman, David C. "Mountaintop Removal Mining: Digging into Community Health Concerns." *Environmental Health Perspectives* 119, no. 11 (November 1, 2011): a476–83.

House, Silas. "My Polluted Kentucky Home." *New York Times,* February 19, 2011, Opinion sec. http://www.nytimes.com/2011/02/20opinion/20House.html?_r=1&scp= 1&sq=silas%20house&st=Search.

———. *Something's Rising: Appalachians Fighting Mountaintop Removal.* Lexington: University Press of Kentucky, 2009.

Howard, Jason. "Appalachia Turns on Itself." *New York Times,* July 8, 2012, Opinion sec. http://www.nytimes.com/2012/07/09/opinion/appalachia-turns-on-itself.html.

Howell, Benita. *Culture, Environment, and Conservation in the Appalachian South.* Urbana: University of Illinois Press, 2002.

Hufford, Mary. "Reclaiming the Commons: Narratives of Progress, Preservation, and Ginseng." In *Culture, Environment, and Conservation in the Appalachian South,* edited by Benita J. Howell, 100–120. Chicago: University of Illinois Press, 2002.

Iafrate, Michael. "Decolonizing Appalachia: Postcolonial Theology in a U.S. American Region." *Catholicanarchy.org,* June 27, 2010. http://catholicanarchy.org/?p=1686.

"iLoveMountains.org—Resolutions of Faith—End Mountaintop Removal Coal Mining." http://ilovemountains.org/resolutions#uua.

James, Thomas. "Responsibility Ethics and Postliberalism: Rereading H. Richard Niebuhr's *The Meaning of Revelation*." *Political Theology* 13, no. 1 (December 4, 2012): 37–59.

Jenkins, Willis. "After Lynn White: Religious Ethics and Environmental Problems." *Journal of Religious Ethics* 37, no. 2 (June 2009): 283–309.

———. *Ecologies of Grace: Environmental Ethics and Christian Theology*. Oxford: Oxford University Press, 2008.

———. *The Future of Ethics: Sustainability, Social Justice, and Religious Creativity*. Washington, DC: Georgetown University Press, 2013.

Jones, Loyal. "Mountain Religion: An Overview." In *Christianity in Appalachia: Profiles in Regional Pluralism*, edited by Bill Leonard, 91–102. Knoxville: University of Tennessee Press, 1999.

Leonard, Bill, ed. *Christianity in Appalachia: Profiles in Regional Pluralism*. Knoxville: University of Tennessee Press, 1999.

Light, Andrew, and Holmes Rolston. "Introduction: Ethics and Environmental Ethics." In *Environmental Ethics: An Anthology*, edited by Andrew Light and Holmes Rolston, 1–11. Blackwell Philosophy Anthologies 19. Malden, MA: Blackwell, 2003.

Loeb, Penny. *Moving Mountains: How One Woman and Her Community Won Justice from Big Coal*. Lexington: University Press of Kentucky, 2007.

MacIvor-Andersen, Josh. "A Brief History of Coal." *Prism: America's Alternative Evangelical Voice* 14, no. 6 (December 2007): 14.

McCauley, Deborah Vansau. "Mountain Holiness." In *Christianity in Appalachia: Profiles in Regional Pluralism*, edited by Bill Leonard, 101–16. Knoxville: University of Tennessee Press, 1999.

McGlynn, Daniel. "Move Not Those Bones." *Sierra* 97, no. 2 (April 2012): 28–33.

McIlmoil, Rory, Evan Hansen, Ted Boettner, and Paul Miller. "Coal and Renewables in Central Appalachia: The Impact of Coal on the West Virginia State Budget." Downstream Strategies and WV Center on Budget and Policy, June 22, 2010.

McNeil, Bryan T. *Combating Mountaintop Removal: New Directions in the Fight against Big Coal*. Urbana: University of Illinois Press, 2011.

Minteer, Ben A., and Robert E. Manning. "Pragmatism in Environmental Ethics: Democracy, Pluralism, and the Management of Nature." In *Environmental Ethics: An Anthology*, edited by Andrew Light and Holmes Rolston, 307–18. Blackwell Philosophy Anthologies 19. Malden, MA: Blackwell, 2003.

Montrie, Chad. *To Save the Land and People: A History of Opposition to Surface Coal Mining in Appalachia*. Chapel Hill: University of North Carolina Press, 2003.

Morrone, Michele, and Geoffrey L. Buckley, eds. *Mountains of Injustice: Social and Environmental Justice in Appalachia*. Athens: Ohio University Press, 2011.

Moyers, Bill. "Moyers on America. Is God Green?" PBS, 2006. http://www.pbs.org/moyers/moyersonamerica/green/index.html.

Murdoch, Iris. *The Sovereignty of Good.* 2nd ed. New York: Routledge, 2001.

Niebuhr, H. Richard. *The Meaning of Revelation.* Louisville, KY: Westminster John Knox Press, 2006.

———. *The Purpose of the Church and Its Ministry: Reflections on the Aims of Theological Education.* New York: Harper, 1956.

———. *Radical Monotheism and Western Culture: With Supplementary Essays.* Library of Theological Ethics. Louisville, KY: Westminster/John Knox Press, 1993.

———. "The Responsibility of the Church for Society." In *The Gospel, the Church and the World,* edited by Kenneth Scott Latourette, 111–32. New York: Harper & Brothers, 1946.

———. *The Responsible Self: An Essay in Christian Moral Philosophy.* New York: Harper & Row, 1963.

———. "War as the Judgment of God." In *War as Crucifixion: Essays on Peace, Violence and Just War from the Christian Century,* edited by John M. Buchanan and David Heim, 17–23. Chicago: Christian Century Press, 2002.

O'Neill, John. "The Varieties of Intrinsic Value." In *Environmental Ethics: An Anthology,* edited by Andrew Light and Holmes Rolston, 131–42. Blackwell Philosophy Anthologies 19. Malden, MA: Blackwell, 2003.

Palmer, M. A., E. S. Bernhardt, W. H. Schlesinger, K. N. Eshleman, E. Foufoula-Georgiou, M. S. Hendryx, A. D. Lemly, et al. "Mountaintop Mining Consequences." *Science* 327, no. 5962 (January 8, 2010): 148–49.

Paulson, Amanda. "In Coal Country, Heat Rises over Latest Method of Mining." *Christian Science Monitor,* January 3, 2006. http://www.csmonitor.com/2006/0103/p02s01-ussc.html.

Peet, Richard, and Michael Watts, eds. *Liberation Ecologies: Environment, Development, Social Movements.* London: Routledge, 1996.

Peterson, Anna L. "Talking the Walk: A Practice-Based Environmental Ethic as Grounds for Hope." In *Ecospirit: Religions and Philosophies for the Earth,* edited by Laurel Kearns and Catherine Keller, 45–62. Transdisciplinary Theological Colloquia. New York: Fordham University Press, 2007.

Presbyterian Church (USA) 217th General Assembly. "Commissioners' Resolution." http://ilovemountains.org/resolutions.

Raney, Bill. "Bill Raney: Congress Must Rein in an Arrogant EPA," December 31, 2010. http://dailymail.com/Opinion/Commentary/201012301192?page=1&build=cache.

Rasmussen, Larry L. *Earth Community, Earth Ethics.* Geneva: WCC Publications, 1996.

Rausch, Fr. John S. "Sowing My Community Back." *Steubenville (OH) Register,* January 31, 2003.

Reclaiming the Future: Reforestation in Appalachia. Lexington: University of Kentucky College of Agriculture, Kentucky State University, 2008. Film.

Reece, Erik. *Lost Mountain: A Year in the Vanishing Wilderness: Radical Strip Mining and the Devastation of Appalachia.* New York: Riverhead Books, 2006.

———. "The Power to Move Perceptions: Orwellian Language in the Land of Coal." In *Plundering Appalachia: The Tragedy of Mountaintop-Removal Coal Mining*, edited by Tom Butler and George Wuerthner, 49. San Rafael, CA: Earth Aware, 2009.

Robbins, Paul. *Political Ecology: A Critical Introduction.* Critical Introductions to Geography. Malden, MA: Blackwell, 2004.

Robertson, David P., and R. Bruce Hull. "Public Ecology: An Environmental Science and Policy for Global Society." *Environmental Science & Policy* 6, no. 5 (October 2003): 399–410.

Ruether, Rosemary R. *Gaia and God: An Ecofeminist Theology of Earth Healing.* San Francisco: HarperOne, 1994.

———. *New Woman, New Earth: Sexist Ideologies and Human Liberation.* New York: Seabury Press, 1975.

Scanlan, Stephen J. "The Theoretical Roots and Sociology of Environmental Justice in Appalachia." In *Mountains of Injustice: Social and Environmental Justice in Appalachia*, edited by Michele Morrone and Geoffrey L. Buckley, 3–31. Athens: Ohio University Press, 2011.

Schweiker, William. *Theological Ethics and Global Dynamics: In the Time of Many Worlds.* Malden, MA: Wiley-Blackwell, 2004.

Scott, Rebecca R. *Removing Mountains: Extracting Nature and Identity in the Appalachian Coalfields.* Minneapolis: University of Minnesota Press, 2010.

Shannon, Thomas R. "The Economy of Appalachia." In *A Handbook to Appalachia: An Introduction to the Region*, edited by Grace Edwards, 67–84. Knoxville: University of Tennessee Press, 2006.

Shapiro, Henry D. *Appalachia on Our Mind: The Southern Mountains and Mountaineers in the American Consciousness, 1870–1920.* Chapel Hill: University of North Carolina Press, 1978.

Shnayerson, Michael. *Coal River.* New York: Farrar, Straus & Giroux, 2008.

Shrader-Frechette, K. S. *Environmental Justice: Creating Equality, Reclaiming Democracy.* Oxford: Oxford University Press, 2002.

Shuford, Chuck. "What Happens When You Don't Own the Land | Daily Yonder | Keep It Rural," July 3, 2009. http://www.dailyyonder.com/what-happens-when-you-dont-own-land/2009/07/03/2205.

Smith, Barbara Ellen. "Legends of the Fall: Contesting Economic History." In *Christianity in Appalachia: Profiles in Regional Pluralism*, edited by Bill Leonard, 1–17. Knoxville: University of Tennessee Press, 1999.

Smith, Peter. "Religion Shaping Mountain-top Removal Debate in Appalachian Coal Country." *Louisville Courier-Journal*, December 19, 2009. http://www.courier-journal.com/article/20091219/NEWS01/912200338/Religion+shaping+mountain-top+removal+debate.

Spadaro, Jack. "Mountaintop Removal: The Destruction of Appalachia." In *Plundering*

Appalachia: The Tragedy of Mountaintop-Removal Coal Mining, edited by Tom Butler and George Wuerthner, 59. San Rafael, CA: Earth Aware, 2009.

Stott, Philip Anthony, and Sean Sullivan. *Political Ecology: Science, Myth and Power*. London: Arnold; Oxford University Press, 2000.

Townes, Emilie M. *Womanist Ethics and the Cultural Production of Evil*. New York: Palgrave Macmillan, 2006.

Unitarian Universalist Association. "End Mountaintop Removal Coal Mining: 2006 Action of Immediate Witness." http://ilovemountains.org/resolutions#uua.

United Methodist Church. "Cease Mountaintop Removal Coal Mining," 2000. http://ilovemountains.org/resolutions.

United States Environmental Protection Agency. *The Effects of Mountaintop Mines and Valley Fills on Aquatic Ecosystems of the Central Appalachian Coalfields (2011 Final)—EPA/600/R-09/138F.* Washington, DC: EPA, 2011.

———. *Mountaintop Mining/Valley Fills in Appalachia Draft Programmatic Environmental Impact Statement—EPA 903-R-00–013.* Philadelphia: EPA, 2003.

Wagner, Melinda Bollar. "Space and Place, Land and Legacy." In *Culture, Environment, and Conservation in the Appalachian South*, edited by Benita J. Howell, 121–32. Chicago: University of Illinois Press, 2002.

Ward, Ken. "Massey Protest Update: 14 Arrested; Accusations Fly," June 18, 2009. http://blogs.wvgazette.com/coaltattoo/2009/06/18/massey-protest-update-14-arrested-accusations-fly/.

Warren, Karen J., and Jim Cheney. "Ecological Feminism and Ecosystem Ecology." In *Environmental Ethics: An Anthology*, edited by Andrew Light and Holmes Rolston, 294–305. Blackwell Philosophy Anthologies 19. Malden, MA: Blackwell, 2003.

West Virginia Coal Association. *West Virginia Coal: Fueling an American Renaissance (Coal Facts 2011)*. Charleston: West Virginia Coal Association, 2011.

West Virginia Council of Churches. "Statement on Mountaintop Removal." Charleston: West Virginia Council of Churches, September 11, 2007.

Weston, Anthony. "Beyond Intrinsic Value: Pragmatism in Environmental Ethics." In *Environmental Ethics: An Anthology*, edited by Andrew Light and Holmes Rolston, 307–18. Blackwell Philosophy Anthologies 19. Malden, MA: Blackwell, 2003.

Whisnant, David E. *All That Is Native & Fine: The Politics of Culture in an American Region*. Chapel Hill: University of North Carolina Press, 1983.

———. *Modernizing the Mountaineer: People, Power, and Planning in Appalachia*. Boone, NC: Appalachian Consortium Press, 1980.

Wickham, James, Petra Bohall Wood, Matthew C. Nicholson, William Jenkins, Daniel Druckenbrod, Glenn W. Suter, Michael P. Strager, Christine Mazzarella, Walter Galloway, and John Amos. "The Overlooked Terrestrial Impacts of Mountaintop Mining." *BioScience* 63, no. 5 (May 1, 2013): 335–48.

Wink, Walter. *The Powers that Be: Theology for a New Millennium*. New York: Doubleday, 1998.

Wuerthner, George. "Appalachia: Land of Diversity." In *Plundering Appalachia: The Tragedy of Mountaintop-Removal Coal Mining*, edited by Tom Butler and George Wuerthner, 3. San Rafael, CA: Earth Aware, 2009.

Yoder, John Howard. *The Politics of Jesus.* Grand Rapids, MI: Eerdmans, 1994.

Zeller, Tom, Jr. "A Battle in West Virginia Mining Country Pits Coal against Wind." *New York Times,* August 14, 2010, Business/Energy & Environment sec. http://www .nytimes.com/2010/08/15/business/energy-environment/15coal.html?_r=2&scp= 2&sq=mountaintop&st=cse.

Index

Place Matters: New Directions in Appalachian Studies
Series Editor: Dwight B. Billings

This series explores the history, social life, and cultures of Appalachia from multidisciplinary, comparative, and global perspectives. Topics include geography, the environment, public policy, political economy, critical regional studies, diversity, social inequality, social movements and activism, migration and immigration, efforts to confront regional stereotypes, literature and the arts, and the ongoing social construction and reimagination of Appalachia. Key goals of the series are to place Appalachian dynamics in the context of global change and to demonstrate that place-based and regional studies still matter.

Appalachia Revisited: New Perspectives on Place, Tradition, and Progress
Edited by William Schumann and Rebecca Adkins Fletcher

The Arthurdale Community School: Education and Reform in Depression Era Appalachia
Sam F. Stack Jr.

Sacred Mountains: A Christian Ethical Approach to Mountaintop Removal
Andrew R. H. Thompson

Rereading Appalachia: Literacy, Place, and Cultural Resistance
Edited by Sara Webb-Sunderhaus and Kim Donehower

CPSIA information can be obtained at www.ICGtesting.com
Printed in the USA
BVOW02*1517281015

424225BV00004B/4/P